Understanding chemistry through microscale practical work

Bob Worley and David Paterson
with Sarah Longshaw

Copyright © The Association for Science Education 2021

First published 2021 by The Association for Science Education.

The Association for Science Education
College Lane
Hatfield
AL10 9AA

The right of Bob Worley and David Paterson to be identified as the authors of this work has been asserted in accordance with the Copyright, Designs and Patent Act 1988.

Photography by Bob Worley and David Paterson
Artwork by Bob Worley and David Paterson
All other images sourced from CLEAPSS

Typesetting and design by Karen Dyer

Printed and bound in Great Britain by Printed Easy, Letchworth.

ISBN 9780863574788

All rights reserved. The material in this publication may be copied for use only within the purchasing organisation. Otherwise, no part of this book may be reprinted or reproduced or transmitted in any form or by any electronic, mechanical, or other means, now known or hereafter invented, including photocopying and recording, or in any information storage or retrieval system without permission in writing from the publishers.

British Library Cataloguing in Publication Data
A catalogue record for this book is available from the British Library

The front cover images show some classic examples of microscale chemistry. From the top left and clockwise: formation of the iron(III)/thiosulfate intermediate, copper hydroxide precipitate, the iron(III)/thiocyanate complex is distorted by a neodymium magnet, citric acid meets sodium carbonate in water with universal indicator and bubbles of carbon dioxide appear, formation of iron(II) hexacyanoferrate(III) (Prussian Blue), sodium chloride solution with universal indicator is electrolysed and pH changes occur at the electrodes.

Dedication

To my family – Claire, Sophie, Ben and Charlie (David Paterson).

To Sue, Matthew and Amelia, Joanna and Chris, and Rosa and Sid (Bob Worley).

To all the science technicians in the United Kingdom who have kept practical science alive despite a heavier workload, reduced hours and a COVID-19 pandemic.

To all teachers of chemistry who value the role of practical work in providing evidence of why chemistry is an essential and exciting subject to their students.

Acknowledgements

Steve Jones, the current Director of CLEAPSS, and Matt Endean (Deputy Director), have been an inspiration in encouraging the authors in the microscale approach over the years. We would also like to acknowledge the help of Magda Polec and Joanna Sacks. On the technical side, Jane Major, Mary Owen and Emma Warwick have all made valuable contributions to the design of special equipment.

Chris Lloyd at SSERC also saw the potential and has spread the ideas to Scotland. The authors are grateful to Chris for reviewing the manuscript on behalf of the ASE Health and Safety Group.

Dr. Julie Hyde, Dr. Peter Hoare and Peter Calder of the Royal Society of Chemistry Secondary and Further Education Group (SaFE) recognised the potential of these techniques as a focus for the Online Conference in July 2021 (https://rscsafegroup.wordpress.com/2021-conference/).

Contents

About this book	1
Chapter 1: Introduction	**7**
Chapter 2: Chemical reactions	**14**
Microscale activity 2.1: Investigating indicators	18
Microscale activity 2.2: Precipitation reactions	19
Microscale activity 2.3: Displacement reactions and reactivity series	20
Microscale activity 2.4: Solid-state displacement reactions	21
Microscale activity 2.5: Solid/gas displacement reactions	22
Chapter 3: Particles, molecules and ions	**24**
Microscale activity 3.1: Diffusion in liquids	27
Microscale activity 3.2: Diffusion of gases	28
Microscale activity 3.3: A microscale alternative to HCl/NH_3 diffusion demonstration	30
Microscale activity 3.4: Diffusion in solutions	30
Chapter 4: Quantitative chemistry – moles and energy	**34**
Microscale activity 4.1: Counting by weighing	37
Microscale activity 4.2: Determining the empirical formula of magnesium oxide	38
Microscale activity 4.3: Determining the percentage water in a hydrated salt	40
Microscale activity 4.4: Determining enthalpy change of reaction	43
Chapter 5: Titration	**46**
Microscale activity 5.1: The mass of a drop of liquid	48
Microscale activity 5.2: Gravimetric titration	49
Microscale activity 5.3: Titration using a syringe-pipette	50
Chapter 6: Rates of reaction and dynamic equilibrium	**53**
Microscale activity 6.1: Investigating the effect of concentration on rate of reaction	58
Microscale activity 6.2: An iodine clock reaction	60
Microscale activity 6.3: Demonstration of conservation of mass	61
Microscale activity 6.4: Determining the equilibrium constant of a redox reaction	62
Chapter 7: Electrochemistry	**64**
Microscale activity 7.1: Conductivity and ions in solution	66
Microscale activity 7.2: Simple chemical cells	67
Microscale activity 7.3: Electrolysis of copper(II) chloride solution	69
Microscale activity 7.4: Conductivity and electrolysis of molten salts	70
Microscale activity 7.5: Determining standard cell potentials	71

Chapter 8: Further microscale activities — 75

Microscale activity 8.1: Identification of nitrate(V) ions — 77
Microscale activity 8.2: Chemistry of iron and other transition metals — 77
Microscale activity 8.3: The chemistry of rusting — 79
Microscale activity 8.4: Catalysis of reaction between zinc and hydrochloric acid — 80
Microscale activity 8.5: Detailed observations of a displacement reaction — 81
Microscale activity 8.6: Liquifying ammonia and the reaction with lithium — 82
Microscale activity 8.7: Synthesis of and equilibrium between interhalogens — 83
Microscale activity 8.8: Catalytic cracking of hydrocarbons — 84
Microscale activity 8.9: Additional polymerisation — 86
Microscale activity 8.10: Synthesis of propene — 87
Microscale activity 8.11: Hydrogenation of propene — 89

Chapter 9: STEM connections — 91

Microscale activity 9.1: Investigating concentration using a DIY colorimeter — 92
Microscale activity 9.2: Investigating the electrolysis of water — 94
Microscale activity 9.3: Chromatography on thin-layer chromatography plates — 96

Chapter 10: Sustainability — 100

Microscale activity 10.1: Investigating sulfur dioxide chemistry — 105
Microscale activity 10.2: Investigating the effects of nitrogen oxides on sulfur dioxide — 106
Microscale activity 10.3: Making esters with heterogeneous catalysts — 107

Postscript from Bob Worley — 109

Figures and Tables

Chapter 1: Introduction — 7
Figure 1.1: The Hickman still — 9
Figure 1.2: The microscale chemistry kit developed at Radmaste — 10
Figure 1.3: Heating iron/sulfur mixtures showing the mineral wool plug — 11

Chapter 2: Chemical reactions — 14
Figure 2.1: Dropper bottles — 15
Figure 2.2: Plastic pipettes — 15
Figure 2.3: An example of a microscale activity sheet held within a plastic wallet — 16
Figure 2.4: A drop of copper(II) sulfate on a polypropylene surface — 16
Figure 2.5: The reaction circles — 18
Figure 2.6: Results of mixed universal indicator in different pH solutions — 19
Figure 2.7: A precipitation reaction worksheet, alongside a reaction drop showing copper hydroxide precipitate — 20
Figure 2.8: Formation of crystals of silver in the reaction between copper and silver nitrate(V) solution taken with a USB microscope — 21
Figure 2.9: Steps in the production of copper by the reduction of copper(II) oxide with carbon — 22
Figure 2.10: The set-up and example of the reaction of copper(II) oxide reduction with hydrogen — 22

Chapter 3: Particles, molecules and ions — 24
Figure 3.1: A Johnstone triangle on the particles, molecules and ions — 25
Table 3.1: The serial dilution of potassium manganate(VII) solution — 26
Figure 3.2: The dissolution and diffusion of potassium manganate(VII) over time — 27
Figure 3.3: A particle diagram of the dissolution of a solute in a solvent — 28
Figure 3.4: Integrated instructions for microscale activity 3.2 — 28
Table 3.2: Some questions and answers when considering diffusion — 29
Figure 3.5: Apparatus set-up for microscale activity 3.3 — 30
Figure 3.6: Formation of silver iodide in a drop — 31
Figure 3.7: Formation of transition metal precipitates — 31
Figure 3.8: A Johnstone triangle on diffusion — 32

Chapter 4: Quantitative chemistry – moles and energy — 34
Figure 4.1: A Johnstone triangle on some aspects of quantitative chemistry — 35
Figure 4.2: Different mass balances available — 38
Figure 4.3: Integrated instructions for microscale activity 4.2 — 39
Figure 4.4: The theoretical masses of magnesium oxide produced vary depending on the empirical formula — 39
Figure 4.5: Construction and use of a bottle top crucible — 40
Figure 4.6: A spirit burner made from a small jam jar — 41

Figure 4.7: Example of energy profile diagrams on the combustion of ethanol   42
Figure 4.8: A Johnstone triangle on some aspects of energy in chemistry   42
Figure 4.9: Integrated instructions for microscale activity 4.3   43
Figure 4.10: Sample data and analysis for microscale activity 4.4   44
Figure 4.11: Digital thermometers   44

Chapter 5: Titration   46
Figure 5.1: The small-scale gravimetric titration set-up   49
Figure 5.2: A microburette   51

Chapter 6: Rates of reaction and dynamic equilibrium   53
Figure 6.1: A Johnstone triangle on rates of reaction   54
Figure 6.2: A reaction progress graph for an exothermic reaction   55
Figure 6.3: A Johnstone triangle on equilibrium   56
Figure 6.4: A reaction profile diagram for a reversible reaction   57
Figure 6.5: An Arrhenius plot of data collected from the acid-thiosulfate reaction at different temperatures   59
Figure 6.6: Small glass vials for microscale reactions   59
Figure 6.7: Reaction boxes used for microscale practicals   60

Chapter 7: Electrochemistry   64
Figure 7.1: A Johnstone triangle on electrochemistry   64
Figure 7.2: Diagrams used to bring together the idea of energy cycles and transfers   65
Figure 7.3: The diffusion of sodium and chloride ions through the puddle, which allows the current to flow indicated by the LED   66
Figure 7.4: The conductivity indicator (the ion detector)   67
Figure 7.5: Measuring the cell potential of a magnesium/ion cell   68
Figure 7.6: The set-up of the microscale electrolysis cell   69
Figure 7.7: A microscale electrolysis apparatus   70
Figure 7.8: The electrolysis set-up for molten salts and production of bromine and lead from lead bromide   71
Figure 7.9: Measuring standard electrode potential of an electrochemical cell   72

Chapter 8: Further microscale activities   75
Figure 8.1: Identification of halide ions with silver nitrate solution   76
Figure 8.2: Nitrate(V) is reduced to ammonia by aluminium in alkaline solution. The ammonia turns the red litmus paper blue and the bromothymol blue from yellow/green to blue   77
Figure 8.3: Reactions of iron(II) and iron(III) ions   78
Figure 8.4: The chemistry of rusting   79
Figure 8.5: An electrochemical cell formed by copper and zinc in acid   80
Figure 8.6: The mechanism and energetics of the reaction of zinc with acid   81
Figure 8.7: The many products of the reaction between magnesium and acidified iron(II) sulfate   82

Figure 8.8: Formation of liquid ammonia and solvated electrons — 83
Figure 8.9: Formation of interhalogen compounds — 84
Figure 8.10: Microscale cracking — 85
Figure 8.11: Synthesis of addition polymers — 87
Figure 8.12: Set-up for producing propene and the smoky flame produced when combusted — 88
Figure 8.13: A three-way tap — 88
Figure 8.14: Reaction of propene and hydrogen over a catalyst — 89
Figure 8.15: Comparison of the flames of propane and propene — 90
Figure 8.16: A catalyst tube — 90

Chapter 9: STEM connections — 91

Figure 9.1: A lateral flow test for SARS-CoV-2 antigens — 92
Figure 9.2: Schematic of the DIY colorimeter, practical set-up using the DIY colorimeter and close correlation between experimental data and Beer-Lambert Law — 93
Figure 9.3: A microscale Hofmann voltameter — 95
Figure 9.4: Microscale chromatography on TLC plates — 97
Figure 9.5: Examples of the effect of a magnetic field — 98

Chapter 10: Sustainability — 100

Figure 10.1: Packed polythene granules at the end of a sealed Pasteur pipette and aluminium oxide, heated with a spirit burner. Alkene gases being collected in an inverted syringe — 104
Table 10.1: Five areas of teaching about sustainable development — 101
Table 10.2: Relevant principles of green chemistry in microscale chemistry — 102
Figure 10.2: Experimental set-up for investigating sulfur dioxide chemistry — 105
Figure 10.3: Experimental set-up for investigating sulfur dioxide chemistry in the presence of nitrogen oxide — 106
Figure 10.4: A sample of a strongly acidic resin used to catalyse esterification reactions — 107

About this book

Microscale chemistry activities offer a way of carrying out practical work that provides many benefits[1]. From adding variety to the practical curriculum, to providing safer ways of achieving results, to shortening practical activities to leave more time for discussion, to reducing costs and waste, these benefits are many and varied.

This publication is designed for teachers and technicians across secondary and further education. Whether you are a student teacher or an experienced Head of Department, whether a new or senior technician, there is something here for you. We hope that the themes will resonate with your practice and that the ideas will inspire further development of your thinking. Perhaps even enthusiastic students might be interested.

The ideas and activities are provided as a starting point. Trial and improvement, experimentation, and just trying something out to see if it works, are as much part of the philosophy of microscale chemistry as laboratory-based research.

The book can be read from cover to cover, or dipped into based on the area of the curriculum that you are teaching or for which you are providing support. Signposts for resources and further reading are included throughout.

A dedicated micro-site is available at www.millgatehouse.co.uk/microscalechemistry where additional ideas and procedures are available.

Safety

Many microscale activities are discussed in this publication. Outline requirements and methods are included as are, in many cases, discussion and images of the outcomes. Often, links are provided in the text to documents containing further activity guidance. Hazards have been identified and risks can be reduced by adopting suitable control measures.

> **Before carrying out any practical activity based on the outline methods presented in this book, readers must fully plan the activity, carry out risk assessments in line with employers' guidelines and rehearse the procedure.**

The UK Health and Safety at Work etc. Act 1974 and subsequent regulations require employers to carry out risk assessments of all activities and all substances being used. Model risk assessments from organisations such as ASE[2], CLEAPSS[3] and SSERC[4] can be used as the basis for the employer's risk assessments. Employees are expected to follow their employer's risk assessments, making suitable adaptation to the local situation of their school and classes.

If you are reading this book in other countries, you must follow your country's or state's Health and Safety Law. Your employer may also have rules that must be followed.

Teachers should:
- check that any fume cupboards used are working correctly;
- check how well ventilated the laboratory is;
- judge whether students are reliable in their behaviour;
- consider whether less hazardous substitutes or lower concentrations work just as well;
- decide whether a demonstration or a class practical would be the best format for the activity;
- make sure that rules on waste are adhered to; and
- rehearse any new activities beforehand so that they are comfortable with the procedure.

One of the main benefits of microscale practical activities is that risks have been naturally reduced by the use of smaller quantities and lower concentrations. In all cases, eye protection (to BS EN166F) should be worn by students and teachers, or goggles (to BS EN166 3) if corrosive or toxic chemicals are being used. Other standards apply in other countries.

Several of the activities in this book are based on CLEAPSS Practical Procedures. Full details are available via the CLEAPSS websites for member schools and colleges. Many equivalent activities are also available via the SSERC website.

Nomenclature

IUPAC rules are generally followed for simple organic compounds, but they can become very unwieldy. Should an 11 year-old use *2 hydroxpropane-1,2,3-tricarboxylic acid* where it is easier and more relevant to them to use *citric acid*? For metals that can exhibit more than two common oxidation states, that will be indicated in the name, thus *copper(II) sulfate 5-water*. We shall not indicate the oxidation state of the sulfate as sulfate(VI). Experience has shown that this can cause great confusion for technicians and teachers from other disciplines. We shall use the term *sulfite* not *sulfate(IV)*.

The units should be expressed as SI units, even though chemists still prefer 'atmospheres' for gas pressures. We have used the symbol M for concentration as shorthand for $mol.dm^{-3}$ or $mol.L^{-1}$

Changing teaching and learning practice

We recognise that changing teaching and learning practices can take time and, in many places, will need a cultural, as well as a technical, change. Ultimately, any change to practice should always be about improving the learning experience for the student. Many years of use of microscale chemistry in our classrooms, and in supporting others to use in their practice, convince us that inclusion of microscale activities is worth the effort.

Joining a community of practitioners can help, and social media provide such platforms. You will receive a warm welcome.

- Twitter – follow @UncleBo80053386, @dave2004b, @CLEAPSS and @SSERCchemistry.
- GoogleGroups – search for, and request to join, the MicroscaleChemistry group.

Bob Worley and David Paterson – October 2021

---

[1] http://science.cleapss.org.uk/Resource-Info/TL018-Microscale-chemistry-poster.aspx [Membership required] (Accessed August 2021)

[2] *Safeguards in the School Laboratory*, 12th Edition, ASE, 2020

[3] http://science.cleapss.org.uk/ CLEAPSS is a subscription organisation in the United Kingdom serving employers in schools and colleges teaching students to 18 years old, in England, Wales and Northern Ireland. Employers will require their teachers and technicians to follow CLEAPSS Guidance. See http://science.cleapss.org.uk/Policies/What-Is-Cleapss.aspx

[4] SSERC covers a similar agenda to CLEAPSS in Scotland. See https://www.sserc.org.uk/

'This is a very clearly written book packed with useful information about how to carry out microscale experiments. Anyone teaching chemistry will benefit from having this in their prep room.'

(Vanessa Kind, Professor of Education, Durham University, UK)

'We often use microscale techniques at our school. We find that students are more confident in practical lessons as it is a much safer way to carry out experiments, and our students get everything they need in one tray (often a takeaway container). This gives more time to discuss observations and scientific concepts. Being adept at microscale techniques has been very useful over the past few years, as it gave our school the confidence that we could carry on with practical work safely during the pandemic. The compact kits could also be used for demonstration under the visualiser without the need to be in a laboratory – we were all very relieved to be able to carry on almost as normal.'

(Sandrine Bouchelkia, Senior Science Technician (research))

'Microscale can be quick, effective, challenging for students, easy to store and easy to prepare. In other words, with little time and students who are making many mistakes because of never having touched lab equipment, it is the perfect way to allow them to efficiently and effectively try again.'

(Chad Hustings, a High School chemistry teacher at Sycamore Community Schools, Mason, Ohio, US. Read his blog article here: https://www.chemedx.org/blog/standards-based-grading-microscale-chemistry-and-post-covid)

A note from Sarah Longshaw, who commissioned and edited this publication

I first encountered microscale activities in a CLEAPSS workshop (at an ASE Annual Conference) led by Bob Worley. His name is often, and in my opinion quite rightly, associated with this way of working. Bob has been an amazing advocate and pioneer of microscale chemistry. I began to research and trial microscale activities in my practice (the simpler ones primarily), as a way to help focus my students' attention and also to reduce quantities of more expensive chemicals (such as silver nitrate). Working with trainees, and also in my role supporting other schools, I have shared various microscale practicals. Trainees find that the procedures help aid behaviour management and I have yet to meet an educator – be that teacher or technician – who doesn't immediately see the benefits of the microscale electrolysis of copper chloride.

I knew David Paterson through his work on integrated instructions before trialling his microscale method for electrochemical cells (one of the A-level practical procedures). David has added his experience from the classroom to this book, as well as an understanding of how to deliver practicals to enhance student understanding. By drawing on the skills of two passionate and experienced practitioners, I believe that we have captured a guide to microscale practical activities and a clear discussion of the pedagogical benefits of the various techniques. Included are suggestions for how to embed this into your practice, as well as some historical context. I hope that you will get as much from reading this guide as I have from being part of its production.

Chapter 1: Introduction

'Gee! In a little I can see a lot'
(an exclamation from an American teacher during a microscale workshop)

Microscale chemistry is a form of practical work that uses much smaller amounts of chemicals and different apparatus and techniques than are often found in UK secondary science laboratories. By reducing scale and simplifying manipulations and observations, students have more capacity to link their practical experience with the theoretical underpinnings of chemistry. In this publication, we make the case that using microscale chemistry practicals can enhance all students' understanding of chemistry. Most of the chapters include an overview of the concepts and common teaching sequences that students will study, along with exemplifications of Johnstone's triangles[1] showing how the concepts are presented in the macroscopic, sub-microscopic and symbolic domains.

Practical work is a fundamental aspect of science education[2], and has a range of uses, including for teaching about scientific enquiry, improving understanding of concepts, teaching practical skills, motivation and to help develop higher level skills, including teamwork and perseverance[3]. The problems with some practical work, particularly 'follow the recipe' style practicals, have been well researched and discussed[4]. Students following step-by-step instructions without much active thought about what they are doing, or why, are not making any conceptual links with their observations and measurements. Indeed, their lack of active thought about the chemistry can be seen from the 'thoughtless' questions[5] that they often ask: *'What do I do next?'*, *'Is this supposed to happen?'*, *'Where is the hydrochloric acid?'*.

The complexity of carrying out multifaceted tasks can quickly lead to an overloaded working memory[6], and making meaning from what they are doing is very challenging for most students. Cognitive load theory[7] has provided some theoretical underpinning for these problems, and can help to inform the decisions that we make about how to design and carry out practical tasks with our students. For example, the *intrinsic* load of carrying out an accurate titration is high due to, amongst other factors, introduction of new equipment, complex skills of controlling the burette tap, swirling the flask and observing the solution simultaneously, as well as assessing the quality of data. The *extraneous* load of a traditional practical session can also be high, since students have to move around the room to collect equipment, negotiate use of shared equipment, and make observations over a wide visual area. When intrinsic and extraneous cognitive loads are high, meaning making and encoding of knowledge into the long-term memory become very difficult.

Microscale chemistry can be used effectively in many ways, reducing or moderating the intrinsic and extraneous loads. For example, by carrying out electrolysis in small drops of water (see Chapter 7), rather than collecting microtubes full of gas, we have simplified the equipment and reduced the extraneous load of using fiddly equipment and the requirement for multiple observations. By carrying out small-scale gravimetric titrations (see Chapter 5) and sequencing the learning steps as we build up to full scale titration, we are reducing the intrinsic load of that full-scale method. Allowing students to draw more on existing knowledge previously gained, rather than being bombarded with many new ideas and observations all at once, allows for better learning and understanding in the long run[8].

Microscale techniques are not new (see next section), but the adoption of microscale chemistry has not been universal across our nations. Some teachers, technicians and departments have been instant 'converts' when they see microscale practicals in action. Others have been more resistant (see the 'Objections' section later in this chapter). In this publication, we hope to make the case for the adoption of microscale chemistry as **part** of the practical curriculum for your students. We do not advocate for complete replacement of 'traditional' practicals with microscale equivalents. Rather, we see microscale as sometimes providing useful alternatives, sometimes providing methods for techniques too dangerous for students at a larger scale, sometimes providing observations that can't be made in other ways and providing a comparison of techniques that can lead to a deeper understanding. To do this, we have combined some theoretical considerations alongside outline methods for microscale techniques.

Many microscale procedures do not require specialist kit. They use standard science equipment such as Petri dishes, dropper bottles and pipettes, but in new contexts. Technicians have an important role to play, not only in dealing with the equipment, but also in enabling the development of innovative ideas. Some kit, such as the micro-electrolysis in a Petri dish, conductivity indicator and the Hofmann Voltameter, need to be made in-house.

Many schools in the UK were sent a copy of the Royal Society of Chemistry (RSC) publication *Microscale Chemistry: Experiments in Miniature* in the 1990s, which contains dozens of practicals. These are now available via the RSC Teach Chemistry website[9], although some methods need to be adapted due to changes in classroom technology (for example, the demise of overhead projectors). A common negative comment at the time was that the microscale activities couldn't be easily demonstrated. However, many teachers now have access to a visualiser (or smartphone or webcams). In fact, when projected onto a large screen, some of the demonstrations can illicit gasps of amazement and illustrate the beauty of chemistry.

The two organisations, CLEAPSS[10] and SSERC[11], also provide many up-to-date trialled practical methods, which are available via the organisations' websites. Many schools and colleges in the UK will be members of these organisations, allowing teachers and technicians to make use of their guidance when planning and risk-assessing practical tasks outlined in this book.

We believe that any teacher or technician with experience in chemistry who is reading this book should be able to create their own procedures and experiments using the ideas gained from this publication. So, while full details are available from other sources, we sincerely hope that you are inspired to experiment with your practice and develop and innovate with the practical work that you use in your classrooms and departments. This enthusiasm and passion for the subject is transmitted to our students, and is seen year after year in the students who we send out into the world[12]. Each chapter includes a 'Teacher activity' section, where further practicals that may be used within the chapter's context can be developed using microscale techniques.

A journey in microscale – a 'little' history from Bob Worley

Microscale chemistry is not new. Techniques were developed from the 19th century onwards, with inorganic analysis pioneered by Friedrich Emich, and microscale quantitative organic analysis by Fritz Pregl (a 1929 Nobel Laureate) at Graz University in Austria. Emich's work fed down into school chemistry in the form of semi-micro qualitative analysis. As part of the three-hour A-level practical exam I took in 1964, students received a small pack with two salts mixed together, to be analysed to determine the metals and radicals present. At university, my research involved analysis of co-polymers, including determining percentage nitrogen in samples using Pregl's techniques and comparison with data from infra-red spectroscopy. In the publication *Practical Chemistry by Micro-Methods* (1929), Egerton Charles Grey applied many of Emich's ideas to teaching medical students in Cairo, including using glass slides for reactions.

Wider development of microscale techniques has been driven by necessity – that being the mother of all inventions. Of particular importance is the need to reduce the cost and environmental impact of chemical experiments, the need to encourage and facilitate practical work in developing nations, and the need to reduce risks for students and teachers during practical work.

Reducing cost of equipment, materials and waste

In a textbook for undergraduate microscale organic procedures[13], the authors note a 75-90% reduction in chemical and disposal cost. Similar cost savings may be possible when microscale activities are used in secondary education. As noted by Dr. Stephen Breuer of Lancaster University, when helping to introduce novel organic practical work with inexpensive equipment to schools in the 1990s, *'Why do we make 5 g of an organic chemical in schools, carry out a melting point analysis and then throw 4.9 g away?'*. This directly inspired the first CLEAPSS Guide on microchemistry techniques. It did use some equipment unfamilar to UK school teachers, such as the Hickman still (Figure 1.1). Semi-microorganic preparation procedures remain a suitable scale, as smaller scale equipment is not always cost effective.

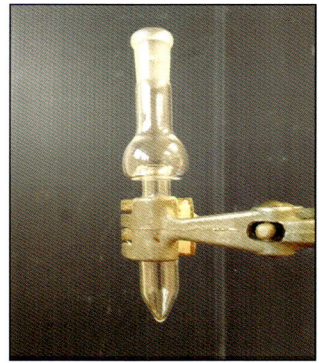

Figure 1.1: The Hickman still

Other equipment can be made in-house, which significantly reduces the cost. For example, making a microscale Hofmann Voltameter has reduced the cost from over £150 to about £20,

Chapter 1: Introduction

most of which is the price of platinum wire. We have even known of schools that have whole class sets. Another saving is using gravimetric titration to introduce students to the technique, to avoid clumsy 14 year-olds breaking burettes and pipettes. CLEAPSS has also designed a colorimeter using LEGO® (see Chapter 9).

Reducing the amount of chemical waste and the associated costs has become increasingly important in the 21st century[14]. With much more emphasis on sustainability in education and chemistry initiated by the Department for Education and the Royal Society of Chemistry respectively, microscale techniques rise in importance.

Encouraging students in developing nations

Stephen Breuer introduced me, Bob, to the 'International Microscale Family' at a meeting in Kuwait in 2011. This community is overwhelmingly friendly and includes practitioners, from high school teachers to university professors from all over the world, with one cause: a love of the microscale approach.

The visit included a presentation on microscale chemistry on Kuwaiti breakfast television alongside Dr. Abdulaziz Alnajjar (former President of the Kuwait Chemical Society) and Dr. John Bradley (Honorary Professor at the University of the Witwatersrand, Johannesburg and founder of the Radmaste institute). John produced a microscale chemistry kit (Figure 1.2), which, with UNESCO funding, was distributed around the world. Schools in the UK were provided with this kit in 1999, alongside the *Experiments in Miniature* book. This kit may well still exist in many schools in the UK. With careful management and replenishment, many practical tasks can still be carried out using the kit.

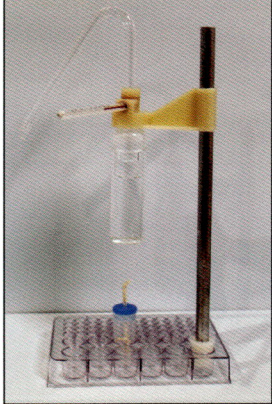

Figure 1.2: The microscale chemistry kit developed at Radmaste

Reducing risk of students and teachers coming to harm

In the early 1990s, CLEAPSS received regular helpline calls about students being taken to hospital due to breathing difficulties. The culprit was the commonly used practical of burning mixtures of iron and sulfur to demonstrate the differences between mixtures and compounds. Diagrams and procedures often showed test tubes half-full of the iron/sulfur mixture. The strong heating of such a set-up resulted in incidents of the sulfur igniting and emitting sulfur dioxide. Employers and safety officers were suggesting that the practical should be withdrawn.

Chapter 1: Introduction

CLEAPSS took a different approach – adapt the experiment to make it safer to use in classrooms. Suitable amounts were calculated using 'Occupational Exposure Limits' set by the UK Health and Safety Executive, and mineral wool plugs were used to prevent the escape of vaporised sulfur. These control measures, as required under the UK Control of Substances Hazardous to Health (COSHH) regulations, allow schools to continue using an invaluable practical. The two methods detailed by CLEAPSS include 2 g of the iron/sulfur mixture[15] (for whole-class demonstration) and 0.2 g for student practicals[16]. The smallest scale benefits even further from working at this microscale. Heating 0.2 g of the mixture in a tube can be carried out using a spirit burner, negating the use of fossil fuels being consumed in raging Bunsen burners, and showing an obvious exothermic reaction (Figure 1.3).

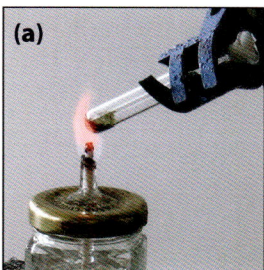 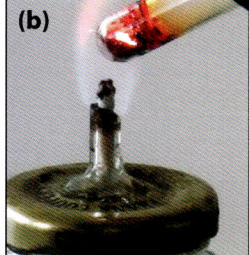 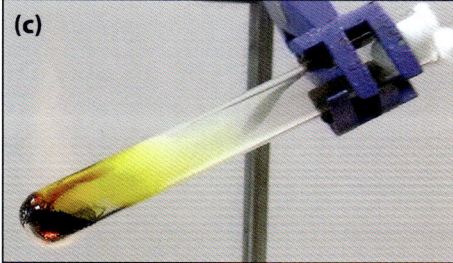

Figure 1.3: Heating iron/sulfur mixtures showing the mineral wool plug. (a) and (b) show the microscale method and (c) the traditional method (Source: CLEAPSS)

Some other common practicals/demonstrations that can be more safely carried out at the microscale include:

- catalytic cracking without suck-back explosions;
- electrolysis of molten lead bromide (toxic) using only 0.5 g on the open bench;
- passing 60 cm^3 of hydrogen over hot copper oxide, without an explosion taking place;
- electrolysis of chloride solutions without producing too much chlorine to affect those with breathing difficulties; and
- studying the rate of reaction (at higher temperatures) of sodium thiosulfate and acid, which can produce a lot of sulfur dioxide.

Objections to microscale chemistry

The kit sent out by the RSC in 1999 seemed totally alien to teachers. It looked like bits of cheap plastic, and many of these kits were placed on shelves to gather dust. Over the course of nearly 30 years of doing microscale and introducing new procedures, we have heard many objections to these ideas, including that:

- students are at a disadvantage in exams if they carry out these new procedures (even though they provide better information and evidence);
- microscale techniques are not mentioned in the examination specification;
- microscale techniques are not mentioned in textbooks and revision manuals;
- 'It is not how I was taught chemistry';
- 'I don't have time to do it and the school does not allow me to go to workshops or ASE conferences';

- quantitative work will not be accurate; and
- 'I have laboratories and the right chemical equipment at school, why should I change?'

John Bradley reports that although the kit has been promoted by UNESCO, it has not revolutionised and improved the teaching of science in developing countries in the way that it was intended to. It is interesting that the reasons that he cited (except the final one) were similar to those experienced in the UK.

The role of exam boards and curriculum developers cannot be underestimated in the promotion of teaching and learning techniques. When traditional practical techniques are used in required/suggested/core practical activities, and replicated in exam questions, these are the techniques that many teachers will use. We are aware that some UK exam boards do promote the use of microscale chemistry[17]. However, until high-stakes exam questions show the use of such techniques, some teachers will remain resistant to using microscale in their general teaching.

Advantages to using microscale chemistry, by David Paterson

I have been using microscale chemistry in my classroom for some time now[18], and both Bob and I have been training teachers and student teachers in microscale techniques over many years. Students are generally very enthusiastic about carrying out practical work, and the speed, efficiency and clarity of observation in many of the microscale practicals really engage the students. Teachers given time to explore the benefits of microscale chemistry in practical sessions tend to be equally enthusiastic, and training sessions are becoming increasingly common in university and school-based training alike. We believe that microscale chemistry can bring significant advantages to all science departments and classrooms. Look out for the following as you read through the book:

- improved safety;
- promotion of practical science;
- addressing environmental and green issues;
- reduced costs in equipment and materials;
- improved classroom management for teachers and technicians;
- improved student focus during practical work (reduced load on working memory);
- more efficient student work, allowing more teaching time;
- improved development of student understanding of difficult chemical concepts by challenging misconceptions about scientific phenomena;
- encouragement of new experiments, e.g. hydrogenation of alkenes and liquefying gases;
- promotion of STEM initiatives with modern materials, e.g. plastics (avoiding single use), carbon-fibre electrodes, Nd magnets; and
- promotion of newly available equipment (digital balances, LEDs, Arduino microelectronics) and modern methods of construction (3D printing, laser cutting, Lego®).

Afterword

And then the pandemic of COVID-19 hit the world, with its impact on science practical work in schools. Would there be no more practicals? Well, many schools undertook microscale practicals even in non-science rooms, just like students in the South African veldt, the Cameroon jungle and Namibian desert.

Roger Ascham wrote a book called *The Scholemaster*, published in 1570, in which he said that gentleness and persuasion, instead of coercion, should be used in schools. Whether you are an experienced chemistry teacher, an early career teacher, a teacher who is teaching out of specialism or a technician supporting practical work within a department, this book is intended to support you and to support your students. Whether you read it from cover to cover, or just dip into a topic as you are preparing to teach it, we hope to gently persuade you to get into the laboratory, try out these techniques, really enjoy practical work, and see chemistry as you may have never seen it before.

'Let the particles do the work.'

References

[1] Taber, K.S. (2013) 'Revisiting the chemistry triplet: drawing upon the nature of chemical knowledge and the psychology of learning to inform chemistry education', *Chem. Educ. Res. Pract.*, 14, 156; https://edu.rsc.org/feature/develop-deeper-understanding-with-models/3010519.article

[2] https://www.gov.uk/government/publications/national-curriculum-in-england-science-programmes-of-study/national-curriculum-in-england-science-programmes-of-study (Accessed August 2021)

[3] https://www.gatsby.org.uk/education/programmes/support-for-practical-science-in-schools (Accessed August 2021)

[4] *Enhancing learning with effective practical science 11-16.* Abrahams & Reiss (Eds.), Bloomsbury, 2016

[5] Johnstone, A.H. (1997) 'Chemistry teaching – science or alchemy', *J Chem Ed*, 74, (3), 262–68

[6] Johnstone, A.H. & Wham, A.J.B. (1982) 'The demands of practical work', *Education in Chemistry*, 71–73

[7] Sweller, J. *et al* (2019) 'Cognitive architecture and instructional design: 20 years later', *Ed Psych Rev*, (31), 261–292

[8] Paterson, D, (2017) 'Making practical work more effective', *Education in Chemistry*: https://edu.rsc.org/feature/making-practical-work-more-effective/3008027.article (Accessed Sept 2021)

[9] https://edu.rsc.org/resources/collections/microscale-chemistry (Accessed August 2021)

[10] http://science.cleapss.org.uk/resources/resource-search.aspx?search=microscale%20chemistry (Accessed August 2021)

[11] https://www.sserc.org.uk/subject-areas/chemistry/chemistry-resources/microscale-chemistry/ (Accessed August 2021)

[12] Sir John Holman always asked first year chemistry students at York University what made them choose chemistry. The answer, year after year, was the enthusiasm and passion for the subject that emanated from their teachers.

[13] Mayo, D.W. *et al* (1986) *Microscale organic laboratory*. John Wiley & Sons

[14] https://microchemuk.weebly.com/2-green-chemistry.html (Accessed August 2021)

[15] http://science.cleapss.org.uk/Resource-Info/PP082-The-reaction-between-iron-and-sulfur-demonstration.aspx [Membership required] (Accessed August 2021)

[16] http://science.cleapss.org.uk/Resource-Info/PP079-The-reaction-between-iron-and-sulfur-Students-version.aspx [Membership required] (Accessed August 2021)

[17] https://www.ocr.org.uk/blog/9-microscale-tips-to-improve-your-chemistry-practical-work/ (Accessed August 2021)

[18] https://edu.rsc.org/endpoint/why-i-love-microscale/4013439.article (Accessed August 2021)

Chapter 2: Chemical reactions

'I remember learning about microscale chemistry practicals at a Royal Society of Chemistry Scholar training day. We learned about the benefits of such an approach. With practicals being an area of teaching I found particularly difficult during my teacher training, the ideas were welcome. Since then, working closely with our wonderful technicians at school, we have tried to incorporate microscale practicals wherever possible. One example includes the microscale displacement of metals using laminated sheets and "drop chemistry". The sheet also includes integrated instructions that have been shown to help students during the practical too. This is a great example because, despite the initial time needed to prepare the sheets and the kits, our technicians understood the benefits in the medium and longer term. With less materials needed we obviously save money. However, it is also the quicker clean-up time that is hugely beneficial for the teacher, students and technicians, which means we would never do this practical any other way'
(Harry Lord, chemistry teacher and Head of Year at Haslingden High School, UK).

Chemical reaction is at the heart of our subject, enabling us to transform substances from one to another, increasing their value and allowing insight into the material world. Society's thoughts on chemical reactions are wrapped up in our experience of our own science education and that of popular culture. From Dr. Frankenstein in Mary Shelley's novel, to Doc Brown in *Back to the Future*, the mad scientist is still the image many have in their mind's eye when thinking of chemistry. The very equipment that we use suggests the image – test tubes and beakers filled with coloured liquids, often bubbling and emitting smoke. Professor Andrea Sella notes the metaphorical meaning of the test tube, seeing it as a place to form ideas and experiment, fundamental in school science.[1]

Carrying out microscale reactions often involves using microlitres of solutions, rather than the usual millilitres. Standard-sized glass test tubes, usually found in schools, are not often suitable for these miniaturised chemical reactions. Since glass is not 100% hydrophobic, a large proportion of a solution can be lost on the glass surfaces when transferring from one container to another. Hence, plastic equipment tends to be used as an alternative material to glass in many of the microscale techniques.

Equipment

For the chemistry described in this book, we do not advocate a specific kit. Throughout the book, examples of equipment will be shown. Most of it will be familiar to teachers and students, but perhaps not traditionally used in chemistry. For example, Petri dishes are commonly used in microbiological experiments, but they also make excellent enclosed reaction vessels for gas reactions in chemistry. Some investment in items such as dropper bottles may be required, but much else tends to already be present in the department's stock. While UK schools are generally well equipped, the issues of replacing items and maintaining them remains.

In this chapter, we will explore some simple chemical reactions that can be carried out on plastic surfaces rather than in test tubes. Basic equipment is first described, then outline details of carrying out the practicals. In later chapters, further equipment is described when used in other practicals.

Dropper bottles

Traditionally, delivering a few drops of reagent is usually achieved with dedicated dropping bottles with a glass pipette and vinyl teat (Figure 2.1a), or from stock bottles using plastic pipettes (Figure 2.1b). Cross-contamination and difficulty in adding drop-wise are common problems with this apparatus. A common alternative used for many microscale activities is the squeeze dropper bottle (Figure 2.1c), familiar to many as eye drop bottles. These are used to deliver the reagent onto the plastic surfaces and spotting tiles. Most aqueous solutions store well. Exceptions are iodine solution (stains the bottle), iron(II) salts (air oxidises), ammonia solution (gas diffuses), hydrogen peroxide (decomposes) and sodium hydroxide solutions (react with atmospheric carbon dioxide). Having said this, only iron(II) salts solutions need to be made *in situ*. The others keep for several months.

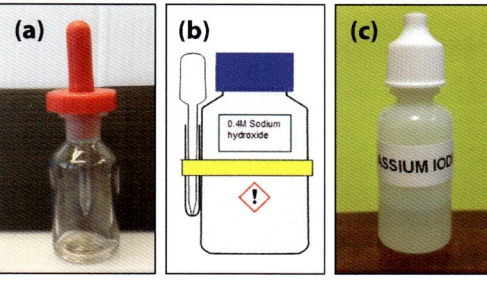

Figure 2.1: Dropper bottles

Pipettes

Plastic pipettes are the workhorses for many of the procedures used in microscale techniques (Figure 2.2).

- The graduated 3 cm^3 pipette (bottom) delivers about 20 drops to release 1 cm^3 of water.
- The graduated 1 cm^3 pipette (centre) delivers about 30 drops to release 1 cm^3 of water.
- The fine tip pipette (top) delivers about 50 drops to release 1 cm^3 of water.

Figure 2.2: Plastic pipettes

These pipettes have been marketed as disposable pipettes, although, with proper care and cleaning, these can be reused many times. There is a further discussion in Chapter 10.

Stirrer

A convenient stirrer can be made from a wooden splint cut to an angled point with scissors. To avoid cross-contamination the pointed end can be re-cut after each stir.

Paper towels

Paper towels can be used to wipe away the solutions after use. Usually these can be placed into the solid waste, but this does depend on the risk assessment.

Plastic surface

Instructions can be printed and inserted in a polypropylene plastic wallet (Figure 2.3). The wallet can then be wiped and reused. The plastic surface is suitable for all aqueous solutions. Issues arise if soap is used, or organic chemicals are added, which reduce the surface tension of the drop. Indicator solutions (which usually contain a mixture of water and ethanol) do work well (see Microscale activity 2.1).

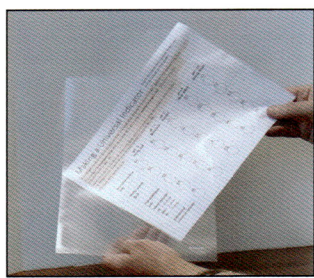

Figure 2.3: An example of a microscale activity sheet held within a plastic wallet

The puddle

Two drops of 0.1 M copper(II) sulfate solution are shown on a polypropylene plastic wallet and on a plastic Petri dish (Figure 2.4). The puddles are hemispheres, with the shape a result of the high surface tension of water combined with the hydrophobic nature of the plastic surface. Chemical reactions and procedures can be carried out in the puddle.

Figure 2.4: A drop of copper(II) sulfate on a polypropylene surface

Making your procedure worksheet

Traditionally, instructions are displayed as a list of equipment and material required, and then a number list of steps, rather like a recipe in a cookbook. This approach is still immensely useful to an experienced student or scientist. However, students at the beginning of their journey into science have many other issues to contend with: the new materials and equipment, the unfamiliar chemical names, the laboratory environment and a myriad worries and thoughts (Am I doing this right? Will it explode? Why does the student next to me keep interfering?), all of which can conspire towards a tentative approach to practical work. Of course, there are other students who will work through the procedure with only a casual regard to the written instructions.

To be effective, a student must have one eye on the instructions and another on the equipment and procedure. This puts quite a load on the student's working memory[2], and there have been significant efforts recently to improve this situation[3,4]. Drawing on the ideas from cognitive science, including the use of dual coding (using both visual and textual information), integrated instruction worksheets are particularly useful in microscale chemistry activities. The co-location of the instruction and the actual chemical reaction provides a particular advantage in reducing the cognitive load of the student.

The printed worksheet with integrated instructions is inserted into a transparent polypropylene wallet (see Figure 2.3). After use, any chemicals on the plastic wallet containing the worksheet can be removed by wiping with a paper towel, and the wallet reused. The printed copy could be laminated, but many teachers adjust worksheets based upon the experience of using them in a previous practical lesson. Disposing of mixed non-recyclable waste should be minimised where possible.

Moving from using test tubes to plastic surfaces is a big change in technique for many teachers. Spotting tiles and well plates can also be used for small-scale reactions; however, procedure and instructions are not as fully integrated.

Laboratory management and PPE

Students can be seated for most microscale procedures since these activities require fine control of dropping bottles or pipettes, so the volumes used are very small and spills are extremely rare. Microscale chemistry provides greater flexibility in how the teacher organises the students carrying out the work. Students can even work on their own, since the quantity/maintenance of equipment is less of an issue. Groups of students can share sets of dropping bottles. Even though most of the solutions are of low hazard, eye protection should still be worn, and other standard safety precautions should be followed.

A common comment about this style of practical is that students cannot cope with the fine control. However, when we consider their use of mobile phones, playing of computer games or even skills in their art lessons, we see that most students can develop good fine motor control.

Demonstrations of the techniques can be carried out using visualisers, webcams or a USB microscope projecting the image to the board, and enable students to see how to carry out the task, and expected results.

Chapter 2: Chemical reactions

Microscale activity 2.1: Investigating indicators

Ensure that full planning and risk assessment is carried out before attempting this activity.

Outline requirements

- dropper bottles of:
  - 1 M ethanoic acid
  - tap water
  - 0.1 M sodium carbonate
  - 0.1 M hydrochloric acid
  - 0.1 M sodium hydroxide
  - pH 4 buffer
  - pH 7 buffer
  - pH 13 buffer (WARNING: Irritant)
- eye protection
- dropper bottles of: (WARNING/DANGER: (Highly) Flammable)
  - methyl orange (MO),
  - bromothymol blue (BB),
  - phenolphthalein (PP),
  - commercial universal indicator (UI).
- instruction worksheets
- plastic wallets
- paper towels

Outline method A – testing the pH of solutions

The technique of working with small drops (or puddles) on a plastic surface works well with indicators and is ideal for the first encounter with microscale techniques. It allows the student to explore the various colours of the indicators with ethanoic acid, sodium carbonate solution and tap water (or pH buffer 7 buffer labelled 'water'). If mistakes are made, the surface can easily be wiped with a paper towel. A worksheet like Figure 2.3 is required. A short 'starter for 10' activity could be introduced with different groups using different indicators and reporting on their findings. Alternatively, once the technique is introduced, students may receive another worksheet with more indicators.

Consider how much washing up is saved. With traditional equipment, the activity would use about 45 test tubes from 15 groups in the class. Adding more indicators only adds to the equipment used.

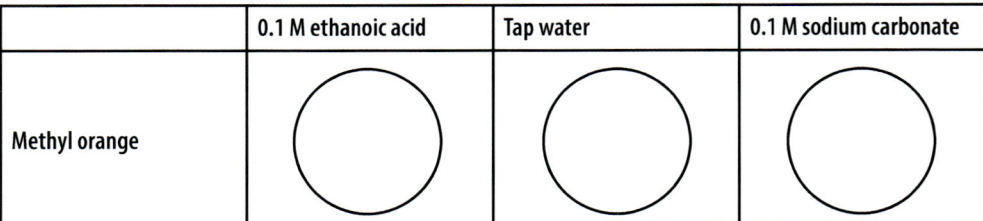

Figure 2.5: The reaction circles

Outline method B – making universal indicator

This is based on CLEAPSS Practical procedure PP057, *Making a universal indicator – a microscale approach*[5].

Using a similar idea, prepare a worksheet with a grid with different pH solutions running along the top, and different indicators running down the side (Figure 2.6). The response of common indicators, such as methyl orange and phenolphthalein, can be investigated by the student. Universal indicator can be made by the students in an empty dropper bottle using 10:5:5 drops of BB:MO:PP, and compared with the commercial indicator. This practical carried out in test tubes would create a huge amount of washing up. Our school technicians are better employed as a valued source of assistance and creativity. Many countries do not have such help.

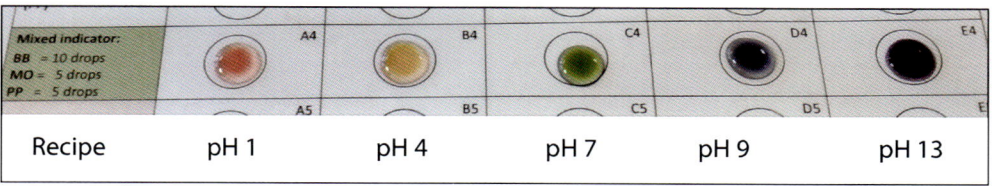

Figure 2.6: Results of mixed universal indicator in different pH solutions

Microscale activity 2.2: Precipitation reactions

Ensure that full planning and risk assessment is carried out before attempting this activity.

Outline requirements

- dropper bottle of 0.1 M salt solutions, e.g.
  - acidified iron(II) sulfate
  - copper(II) sulfate
  - iron(III) nitrate
- dropper bottle of 0.4 M sodium hydroxide (WARNING: Irritant)
- eye protection
- paper towels
- reaction worksheet in a plastic wallet

Outline method

Seeing precipitates suddenly appear can be magical. Two liquids are mixed and suddenly a solid appears. Sometimes the solid is a different colour to the solutions. When copper sulfate solution is mixed with sodium hydroxide, a blue solid appears, as the insoluble copper hydroxide is formed.

Figure 2.7 shows an example of a worksheet for a precipitation reaction, alongside how the precipitate appears after stirring. The square has a shaded part so that white precipitates can be seen better. This method can be applied to any mixture of aqueous solutions that, on mixing, produce an insoluble product. This technique is used for identifying metals and non-metals in salts.

Chapter 2: Chemical reactions

The Copper Hydroxide Precipitate

Laminate the paper/insert the paper into a plastic (eg, polypropylene) wallet. Wear eye protection.

1. Add 2 drops of 0.1 – 0.2M copper sulfate solution to the square
2. Add 2 drops of 0.4M sodium hydroxide on the solution
3. Stir mixture with a with wooden splint and then cut off the end.
4. (optional) Leave the puddle for 30 minutes and see if there are any changes.

Figure 2.7: A precipitation reaction worksheet, alongside a reaction drop showing copper hydroxide precipitate

Microscale activity 2.3: Displacement reactions and reactivity series

Ensure that full planning and risk assessment is carried out before attempting this activity.

Outline requirements

- Small pieces of magnesium, copper, iron, zinc
- dropper bottle of:
  - 0.05 M silver nitrate(V) solution (Can stain skin)
  - 2 M ammonia (DANGER: Corrosive, irritant)
  - 0.1 M magnesium sulfate
- 0.1 M copper(II) sulfate
- 0.1 M iron(II) sulfate [freshly prepared]
- 0.1 M zinc sulfate (WARNING: Irritant)
- goggles
- instruction sheet in plastic wallet
- paper towels

Outline method

The reaction between silver nitrate and copper makes a spectacular start to these investigations (Figure 2.8).

- Photocopy the page or draw a 1.5 cm diameter circle on paper.
- Insert the copy or paper into a plastic wallet.
- Add about 3 to 5 drops of 0.05 M silver nitrate solution.
- Move a piece of bare copper wire (or small pieces of thin foil) into the puddle. Watch what happens over the next 5 minutes.
- When the reaction is complete, remove the copper wire and solid that adheres to it and add 2 drops of 2 M ammonia solution.

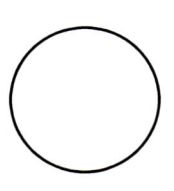

Small-scale reactions between metals and metal solutions are commonly carried out in spotting tiles or well plates. Carrying out the procedure in puddles has some visual and teaching advantages. Useful metals to investigate are magnesium, zinc, iron and copper. The chemistry is much more complicated than textbooks would have us believe. The keen student will see bubbles of gas appear (hydrogen) and possibly precipitates (hydroxides). Metal chlorides and sulfates can be used as salts but beware of using nitrates, which can render metals passive. Iron(II) salts are a particular problem since they are oxidised quickly by atmospheric oxygen. It is best to make the solution just before it is used. (If acid is added, it only complicates the observations.)

Figure 2.8: Formation of crystals of silver in the reaction between copper and silver nitrate(V) solution taken with a USB microscope

Microscale activity 2.4: Solid-state displacement reactions

Ensure that full planning and risk assessment is carried out before attempting this activity.

This activity is based on CLEAPSS Practical Procedure PP077, *Metal oxide reduction with carbon*[6].

Iron (steel framework), copper (electrical wiring), zinc (batteries) and aluminium (food containers) are encountered by students every day. Extracting the metals from the ores is a vital part of the modern economy, usually with great energy and environmental costs. However, extracting these elements in the laboratory is a great experience for the student, and has traditionally been carried out using porcelain crucibles. Porcelain crucibles are not really designed for multiple uses, and can break during the strong heating required.

Outline requirements

- carbon
- copper(II) oxide (WARNING: Harmful)
- beaker of tap water
- bottle top crucible (see Chapter 4)
- Bunsen burner
- eye protection
- clay triangle
- conductivity meter (see Chapter 7)
- heat-proof mat
- spatulas
- tongs/pliers
- tripod

Outline method

1. Mix together 0.5 g of copper(II) oxide and 0.2 g of carbon on two pieces of folded paper.
2. Mix by adding one powder to the other powder and back again several times to ensure good mixing.
3. Set up the crucible on a pipe clay triangle on a tripod.
4. Heat the mixture strongly until the reaction is complete (Figure 2.9).

Chapter 2: Chemical reactions

Figure 2.9: Steps in the production of copper by the reduction of copper(II) oxide with carbon (Source: CLEAPSS)

5. Allow the crucible to cool for a minute, then hold on the surface of water with the tongs/pliers to cool the crucible and reaction mixture.
6. Test the conductivity of the red product.

Microscale activity 2.5: Solid/gas displacement reactions

Ensure that full planning and risk assessment is carried out before attempting this activity.

This activity is based on CLEAPSS Practical Procedure PP077, *Metal oxide reduction with hydrogen*[7].

Reduction of copper(II) oxide with hydrogen can be carried out at large scale. However, there is the chance of a hydrogen/air mixture forming, which, if ignited, can cause an explosion. Indeed, a teacher was prosecuted under the Health and Safety at Work act in 1984 for endangering students when such an experiment failed. Following this incident, CLEAPSS developed a smaller-scale method, originally based on the Radmaste Kit. In this set-up (Figure 2.10), hydrogen is produced *in situ* by the reaction of zinc and sulfuric acid. The hydrogen passes over copper(II) oxide in a Pasteur pipette, and is heated by a spirit burner. The oxidation of copper(II) oxide can be clearly seen, and the formation of water as a second product can also be observed.

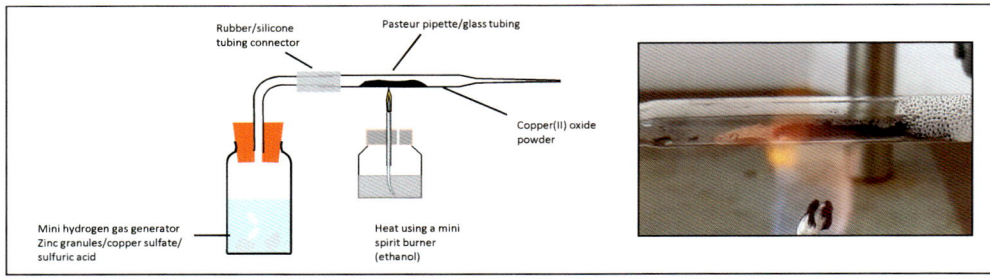

Figure 2.10: The set-up and example of the reaction of copper(II) oxide reduction with hydrogen (Source: CLEAPSS)

Teacher activities

- Consider how you could carry out reactions, which you have usually done in test tubes, in a small drop. Consider which year groups you would introduce microscale practicals to.
- How might you investigate indicators other than the commercially available ones? For example, many plants provide useful natural indicators (e.g. azalea, rose and pansy flower heads, black grape and blueberry skins, and poinsettia and red cabbage leaves, see Further Reading).
- Cross-curricular investigations: artists have used colour in works of art for millennia. They made the colour from pigments (ground-up insoluble inorganic chemicals) and dyes (organic chemicals) found in flowers or vegetables. The old artists made synthetic pigments, such as Prussian Blue and French Ultramarine, to replace expensive mineral lapis lazuli. In the 19th century, chemists developed synthetic dyes. Microscale chemistry can be used to produce fascinating artworks (see back cover).

Further reading

Aqueous red cabbage extracts: More than just a pH indicator, which can found at https://www.chemedx.org/article/aqueous-red-cabbage-extracts-more-just-ph-indicator Iain A. Smellie, Iain L. J. Patterson, Adrian Allan, Bob Worley. Although this is an American website, the authors are Scottish with one exception.

References

[1] https://thebiomedicalscientist.net/science/test-tube-symbolic-story (Accessed August 2021)

[2] Johnston, A.H. & Wham, A.J.B. (1982) 'The demands of practical work', *Education in Chemistry*

[3] Paterson, D.J. (2019) 'Design and Evaluation of Integrated Instructions in Secondary-Level Chemistry Practical Work', *J. Chem. Educ.*, **96**, (11), 2510–2517. See https://dave2004b.wordpress.com/integrated-instructions for more details (Accessed August 2021).

[4] https://edu.rsc.org/feature/improving-practical-work-with-integrated-instructions/3009798.article (Accessed August 2021)

[5] http://science.cleapss.org.uk/Resource-Info/PP057-Making-a-Universal-Indicator-A-Microscale-Approach.aspx [Membership required] (Accessed August 2021)

[6] http://science.cleapss.org.uk/Resource-Info/PP077-Metal-oxide-reduction-with-carbon.aspx [Membership required] (Accessed August 2021)

[7] http://science.cleapss.org.uk/Resource-Info/PP076-Microscale-reduction-of-copper-II-oxide-with-hydrogen-gas.aspx [Membership required] (Accessed August 2021)

Chapter 3: Particles, molecules and ions

'Although it has been around for decades, it is only in the last three years that I have really embraced the use of microscale chemistry. Aside from the pedagogical and cost benefits, I have found that the use of microscale and reduced scale has reignited an interest in practical work within my pupils. Whereas, it may have been the case that the practical work was a "break from the theory", pupils now are excited about watching, for example, how two crystals of soluble salts in a small water "puddle" diffuse together to create, through a swirling pattern, a precipitate. When we have the time, I encourage the pupils to film the whole process in time-lapsed photography using their smartphones. The pupils find seeing the whole process at this speed fascinating and I have to say that I'm equally excited!'

(Jonathan Peverley, Head of STEM, Oundle School, UK)

Concepts being developed

If, as the European Space Agency suggests[1], there are 10^{11}-10^{12} stars in our galaxy, and 10^{11}-10^{12} galaxies in our universe, then the number of stars in the universe is between 10^{22} and 10^{24}. The fact that there are a similar number of water molecules (6×10^{23}) in 18 cm^3 of water is an extremely challenging idea for students, but a critical *big idea*[2] of the particulate nature of matter.

A key problem with teaching chemistry is convincing students (and most of the general public) that these particles exist, especially as they cannot be seen. The late Alex Johnstone (AHJ) summed up the difficulties in teaching chemistry in his famous triplet[3], later shown as a triangle[4]. It is comforting to us all that such a well-respected teacher and later education researcher such as AHJ realised just how difficult it was to teach chemistry (Figure 3.1).

The macro events observed and experienced by the student can be interpreted in two ways:

Chapter 3: Particles, molecules and ions

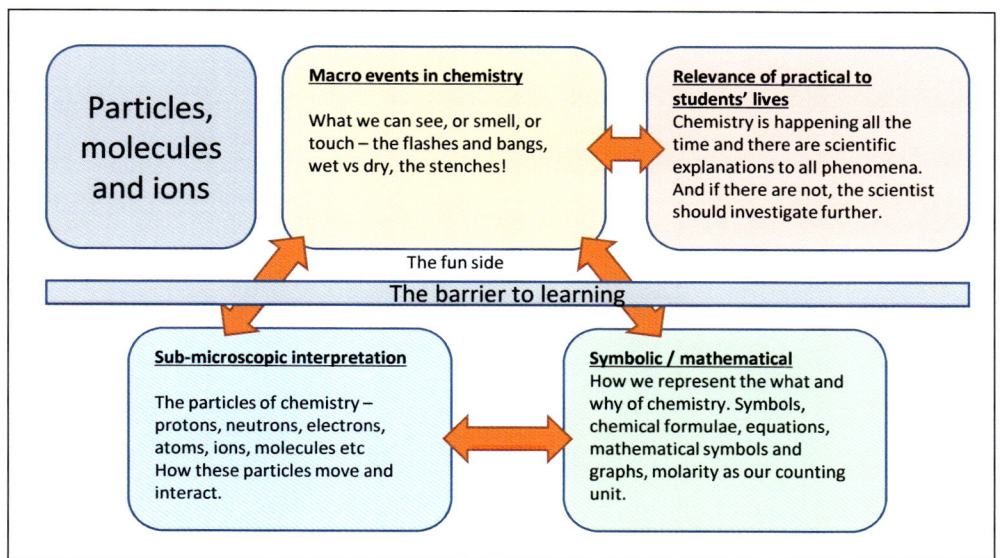

Figure 3.1: A Johnstone triangle on the particles, molecules and ions

- By using chemical formulae and chemical equations as a form of shorthand to describe the events. Using mathematical equations, relating to energy changes, rate of reaction and equilibrium etc.; and
- By using models, through diagrams and physical models (e.g. sticks and balls) of the particles, we visualise these events, creating mental models of the chemical activity. Even though the circles and spheres that we use as a representation of the actual particles are ultimately inaccurate, this process is useful.

In doing this, AHJ realised that teachers should not overload the limited working memory capacity of their students. We should not attempt to tackle all three corners of the triangle in one grand swoop, rather help the student to understand each vertex and be explicit about how they relate to each other.

AHJ also realised that relevance was important. Students need to appreciate that there are chemical events happening inside them (e.g. respiration) and around them (e.g. using fuel, cooking, washing) every moment of the day. Some authors have turned the triangle into a tetrahedron, but that is difficult to show in 2 dimensions along with comments.

Chemistry is difficult when we cross the barrier between the observable (macro) to the abstract (sub-micro and representational). Hard work and thinking is required to understand. Confusion is common, and often demotivational, but is a vital part of the learning process. So, although younger students often love chemistry because of the macro events, it rapidly becomes a 'hard' subject as we delve down into the sub-microscopic world. This is exemplified by one response from a former student: *'Bob, I really enjoyed all the experiments you showed us and loved the "turd" reaction (sugar and concentrated sulfuric acid), but I could not do the equations or calculations'*.

A little goes a long way – detecting the presence of ions

Potassium manganate(VII) is an intensely coloured compound, a useful property should we wish to show how far we can divide up a few crystals and yet still see the colour. Table 3.1 shows the serial dilution of a 1% solution of potassium manganate(VII). The particles in potassium manganate(VII) carry an electric charge, which allows an electric current to flow through the solution. A conductivity meter was used to measure the conductivity of the sixth jar at 33 μScm^{-1}. The reading for a sample of pure water was 3 μScm^{-1} The meter detects the presence of these particles although our eyes cannot.

Mass of $KMnO_4$ in 10 cm³ / g	Concentration of $KMnO_4$ / M	Particles of $KMnO_4$ in the pot	Solution
0.1	0.063	3.8×10^{22}	
0.01	0.006 3	3.8×10^{21}	
0.001	0.000 63	3.8×10^{20}	
0.000 1	0.000 063	3.8×10^{19}	
0.000 01	0.000 006 3	3.8×10^{18}	
0.000 001	0.000 000 63	3.8×10^{17}	

Table 3.1: The serial dilution of potassium manganate(VII) solution

Diffusion is the process of different substances mixing due to the random movement of the particles. In gases, since all components are fully miscible, a near uniform mixture is achieved as the gases each fill the available volume. Diffusion of solutes through solvents is much slower, but eventually a solution of uniform concentration is achieved.

Chapter 3: Particles, molecules and ions

When substances react in air or in solution, products will form. If the products have different properties from the original substances, observations can be made. These observations are useful evidence for the dissolution and diffusion of particles.

Microscale activity 3.1: Diffusion in liquids

Ensure that full planning and risk assessment is carried out before attempting this activity.

This is based on an experiment in Alex Johnstone's *Chemistry Takes Shape, Book 1* (1964).

Outline requirements

- potassium manganate(VII) (a few crystals) (DANGER: Oxidiser, harmful)
- dropper bottle of water
- eye protection
- wooden splint
- plastic wallet and paper
- smartphone with timelapse camera (optional)

Outline method

1. Draw a 1.5 cm diameter circle on paper.
2. Insert the paper into a plastic wallet.
3. Make a puddle of water about 1.5 cm across over the circle.
4. Place a couple of crystals about 1 cm from the puddle.
5. Carefully push a couple of crystals of potassium manganate(VII) into the edge of the puddle.
6. Observe the mixture over time. A timelapse camera can be used to follow the diffusion over time (Figure 3.2).

Example results

| Start | 2 min | 4 min | 8 min | 16 min | 32 min |

Figure 3.2: The dissolution and diffusion of potassium manganate(VII) over time

The action of diffusion can be modelled using simple diagrams (Figure 3.3) showing the potassium manganate particles moving through the water particles. Multiple representations of the particles moving chaotically (verbally, visually, diagrammatically, etc.) will help students to accept the existence of these sub-microscopic and individually invisible particles.

Many gases with interesting chemistry, for example ammonia, chlorine, hydrogen sulfide and sulfur dioxide, are harmful or toxic. Class practicals that generate these gases can cause significant risks for students and the inhalation of these gases has sent students to hospital.

Chapter 3: Particles, molecules and ions

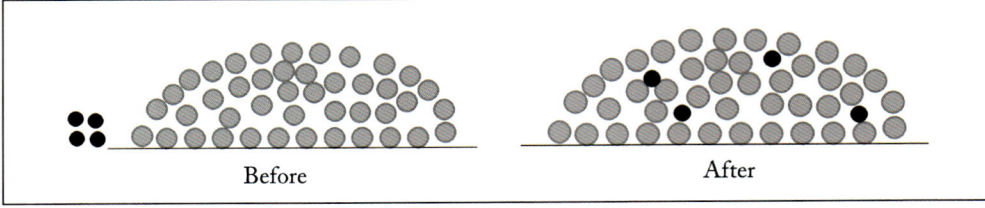

Before   After

Figure 3.3: A particle diagram of the dissolution of a solute in a solvent

However, by carrying out the practicals on a microscale, inside a Petri dish, much lower volumes of the gases are produced and they are mostly contained. We will still be able to smell these gases, so well-ventilated classrooms are important!

Microscale activity 3.2: Diffusion of gases

Ensure that full planning and risk assessment is carried out before attempting this activity.

This is based on an activity originally from the RSC *Microscale chemistry* publication – *diffusion of gases on a microscale*[5].

Outline requirements

- dropper bottles of
- 0.1 M potassium iodide
- 1% starch
- 50% domestic chlorine-based thin bleach (WARNING: Irritant)

- 1 M hydrochloric acid
- eye protection
- instruction sheet in plastic wallet
- paper towel
- Petri dish lid

Outline method

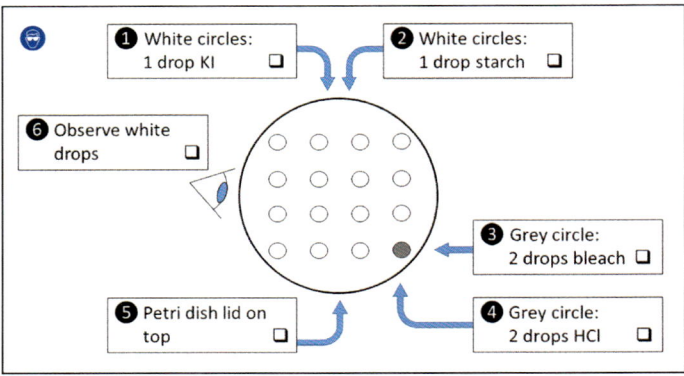

Figure 3.4: Integrated instructions for microscale activity 3.2

1. Insert the worksheet into a plastic wallet.
2. Place one drop each of potassium iodide and starch on each of the white circles.
3. Place two drops of bleach then two drops of acid on the grey circle.
4. Quickly place the Petri dish lid over the drops.
5. Observe the white drop.

28   Understanding chemistry through microscale practical work

The chlorine is contained within the enclosed Petri dish, and the diffusion is observed by the white drops closest to the grey drop turning blue/black first. The enclosed reaction prevents draughts so clear results are easily obtainable. If the set-up goes wrong, it is quick and easy to wipe down the plastic wallet and start again.

Discussing diffusion

It can be tempting to try to explain diffusion 'as occurring when gases want to spread around the room to ensure the concentration is the same throughout'[6]. It is better perhaps to discuss based on questions and answers (Table 3.2).

Question	Answer	Comment
What is the approximate diameter of a particle?	They start at 10^{-10} m and can get larger	Electron microscopes are just able to show particles. Electron and X-ray diffraction patterns, caused by the arrangement of particles in solids (lattices), have been known for some time.
What is the distance between these particles in a gas?	Between 10^{-8} and 10^{-9} m	If the diameter of the particle was 1 cm, the next particle would be 10 cm away (on average). There is nothing in between.
How fast are these particles moving in a gas at room temperature and pressure?	About 500 ms^{-1} for nitrogen at room temperature	This is 1800 km hr^{-1} or 1100 miles hr^{-1}.
How often do particles collide?	Between 10^9 and 10^{10} collisions s^{-1}	The movement of the particles is random. When chemicals are mixed, some collisions might result in a chemical reaction.

Table 3.2: Some questions and answers when considering diffusion

This means that:
- the particles of a gas (and in any matter) are extremely small;
- there is a lot of space between the particles in a gas, but less in a liquid and a solid;
- the particles are moving very quickly; and
- there are countless collisions (chaos) resulting in the transfer of energy between the particles.

It is the random movement and collisions that cause the particles to disperse throughout the room in an apparently orderly fashion. The explanation of order from chaos seems counterintuitive and not 'common sense'.

Diffusion occurs in liquids, but the progress is slower as shown by the potassium manganate(VII) dissolving in a puddle. Diffusion is possible between solids. If a clean sodium surface is put into contact with a clean potassium surface, liquid sodium potassium alloy is formed. If copper is zinc-plated (grey colour) and heated, a yellow-gold coating appears as a thin layer of brass (copper/zinc alloy) forms[7].

Chapter 3: Particles, molecules and ions

Microscale activity 3.3: A microscale alternative to HCl/NH$_3$ diffusion demonstration

Ensure that full planning and risk assessment is carried out before attempting this activity.

Outline requirements

- concentrated hydrochloric acid
  (DANGER: Corrosive)
- concentrated ammonia solution
  (DANGER: Corrosive)
(both need to be recently purchased or well-stored stock)

- goggles
- glass Pasteur pipettes (2)
- silicon tubing
- clamp, boss and stand

Outline method

1. Ensure that the room is well ventilated and the activity has been rehearsed.
2. Connect the two pipettes with a short piece of silicone tubing (see Figure 3.5).
3. Dip one pipette tip in the hydrochloric acid and the other in the ammonia solution.
4. Clamp the set-up horizontally.
5. Observe for about 3 minutes – a ring of ammonium chloride will appear closer to the acid end.

Measuring the distance of the disc from either end illustrates that the lighter gas has travelled further, in the same time period, than the heavier gas. There are billions of air particles in the way of the hydrogen chloride and ammonia particles, but these particles still get through the air particles to react – chaotic random motion somehow produces order.

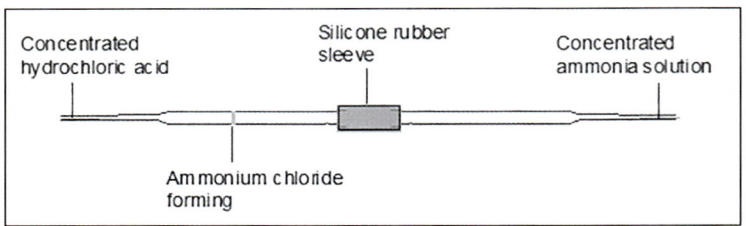

Figure 3.5: Apparatus set-up for microscale activity 3.3

Microscale activity 3.4: Diffusion in solutions

Ensure that full planning and risk assessment is carried out before attempting this activity.

'We did precipitates last year' is a comment heard from many students every year. Re-covering concepts over the course of a student's' chemical education is important – the spiral curriculum. However, there should be new ideas and observations each time to help develop their understanding.

Many of the solutions that students used are already made up for them. They appear as 'water', given that they are often clear and colourless. The sudden appearance of a solid can appear as 'magic'. This activity combines dissolving, diffusion and precipitation.

Chapter 3: Particles, molecules and ions

Outline requirements

- potassium iodide (a few crystals)
- silver nitrate(V) (a few crystals) (DANGER: Oxidiser, corrosive)
- eye protection
- dropper bottle of water
- piece of paper
- plastic wallet
- wooden splint

Outline method

1. Make a puddle of water about 1 cm across (15-20 drops).
2. Carefully place a few crystals of potassium iodide in one side of the drop, and a few crystals of silver nitrate in the other side.
   a. Either push the crystals into the drop using a dry splint, or
   b. dampen the end of the splint, dip into a pile of the crystals and then dip into the drop.
3. Observe the drops over time (Figure 3.6 and Figure 3.7).

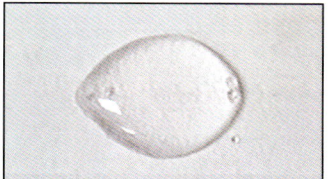  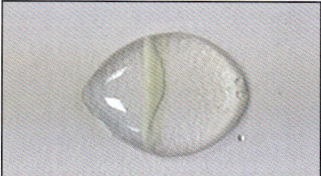

Figure 3.6: Formation of silver iodide in a drop

Example outcome

Many other precipitation reactions are possible (Figure 3.7)

copper(II) carbonate	iron(II) carbonate	lead(II) iodide[8]	iron(II) hexacyanoferrate(III)

Figure 3.7: Formation of transition metal precipitates

By demonstrating the existence of ions in solids, that ions exist in water when salts dissolve and that ions move in solutions, we can discuss the formation of compounds in reaction mixtures. Discussing the attraction between ions and water molecules, in the contexts of these microscale results, helps to make sense of the chemical equations that we wish our students to fully appreciate:

$$KI(aq) + AgNO_3(aq) \rightarrow AgI(s) + KNO_3(aq)$$

Diffusion is a vital concept for students to understand in chemistry (Figure 3.8), as well as in biology and physics. The advantage of reactions in these puddles of solution is that the observations are so clear that discussion and understanding can be built on these observations. When this concept is reinforced time and time again, gradually the student will grow in their understanding and accept the particulate nature of matter.

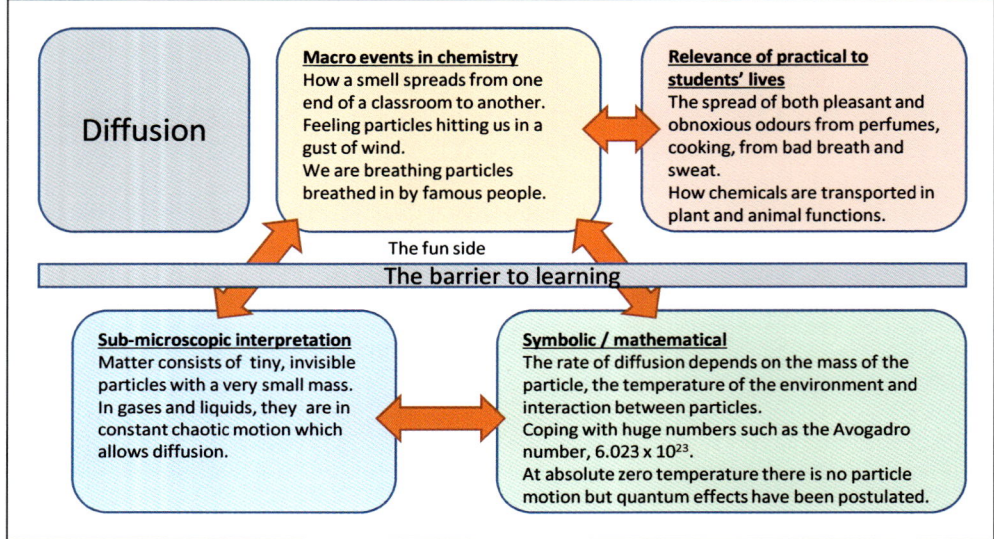

Figure 3.8: A Johnstone triangle on diffusion

Teacher activities

- Consider how you would structure a series of activities to help students understand the presence of particles that they can't see, or the process of precipitation.
- Consider how the reactions of ammonia could be investigated using Petri dish containers. Many transition metal salt solutions will react with ammonia, some producing interesting halo effects within the drop.
- Consider how the reactions of acidic gases could be investigated using Petri dish containers. Sulfur dioxide and nitrogen oxides can be produced on the microscale and react differently with oxidising and reducing agents.
- Consider drawing your own Johnstone's triangles when you are planning for a curriculum topic. How will this help with your own conceptualisation of the chemical ideas?

Further reading

Bob Worley *et al* (2019) 'Visualizing Dissolution, Ion Mobility, and Precipitation through a Low-Cost, Rapid-Reaction Activity Introducing Microscale Precipitation Chemistry', *J. Chem. Educ.*, **96**, (5), 951–954. https://doi.org/10.1021/acs.jchemed.8b00563. Teachers can access this by going to https://www.chemedx.org/article/jce-9605-may-2019-issue-highlights (Accessed October 2021)

References

[1] https://www.esa.int/Science_Exploration/Space_Science/Herschel/How_many_stars_are_there_in_the_Universe (Accessed August 2021)

[2] https://www.ase.org.uk/bigideas (Accessed August 2021)

[3] Johnstone, A.H. (1982) 'Macro- and microchemistry [Notes and correspondence]', *Sch. Sci. Rev.* **64**, (227), 377–379

[4] Taber, K.S. (2013) 'Revisiting the chemistry triplet: drawing upon the nature of chemical knowledge and the psychology of learning to inform chemistry education', *Chem. Educ. Res. Pract.,* **14**, 156

[5] https://edu.rsc.org/experiments/diffusion-of-gases-on-a-microscale/535.article (Accessed August 2021); Paterson, D. (2019) 'States of matter and particle theory', *Education in Chemistry,* https://edu.rsc.org/cpd/states-of-matter-and-particle-theory/3010239.article (Accessed August 2021)

[6] We need to be careful in attributing human characteristics to inanimate objects such as these particles (anthropomorphism). We should be careful of teleological explanations where the knowledge of the end result drives the explanation of the mechanism. It is a difficult and controversial area of discussion.

[7] A quoted reaction between solid lead nitrate and solid potassium iodide producing yellow lead iodide actually occurs because there are trace amounts of water present. Drying the chemicals at 80°C in an oven before mixing produces no colour change.

[8] This is the only precipitation that forms crystals that students can see glistening in the light. Using the USB microscope shows beautiful effects caused by the bombardment of water molecules on the tiny crystals.

Chapter 4: Quantitative chemistry – moles and energy

'Microscale practicals help me to demonstrate core concepts in a much shorter timeframe, whilst still allowing my students to see the chemistry for themselves. The conservation of mass experiment allows students to construct their own explanations of why masses remain constant in chemical reactions, and why it appears to change when one product is a gas. I have always found this to cement their understanding effectively, and it's always an integral part of my early GCSE teaching!'
(Louise Hussein, chemistry teacher in the UK).

Concepts being developed 1 – moles

Quantitative chemistry brings together the ideas of macroscopic handling of substances in the laboratory, microscopic consideration of what is happening to the particles in a reaction, and the symbolic representation of chemical reactions through chemical and numerical equations. With so many representations, it is little wonder that this can be one of the hardest topics for students to engage with and master (Figure 4.1).

Learning quantitative chemistry effectively requires constant review and use of many different contexts. This will help students to become comfortable with manipulating models of invisible particles in their minds, and relating observations in the real world to our theories of the working of the microscopic world. There needs to be plenty of use exemplification of the equations involved and plenty of practice, spread out over the curriculum. This repeated, integrated practice should help embed the central knowledge and skills in the students' minds. For those moving on to post-16 chemistry study, a solid grasp of quantitative chemistry is critical to success.

Chapter 4: Quantitative chemistry – moles and energy

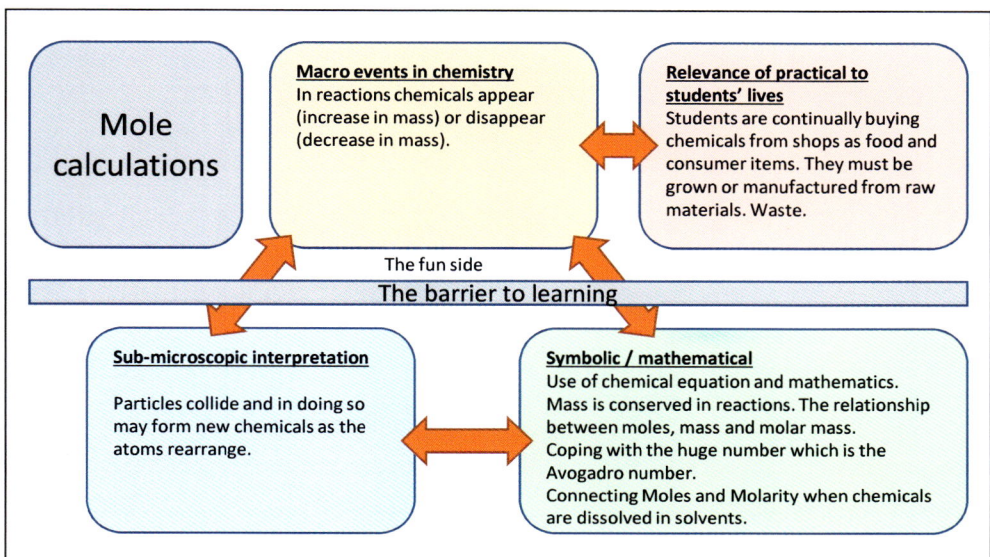

Figure 4.1: A Johnstone triangle on some aspects of quantitative chemistry

A possible teaching sequence for quantitative chemistry is :

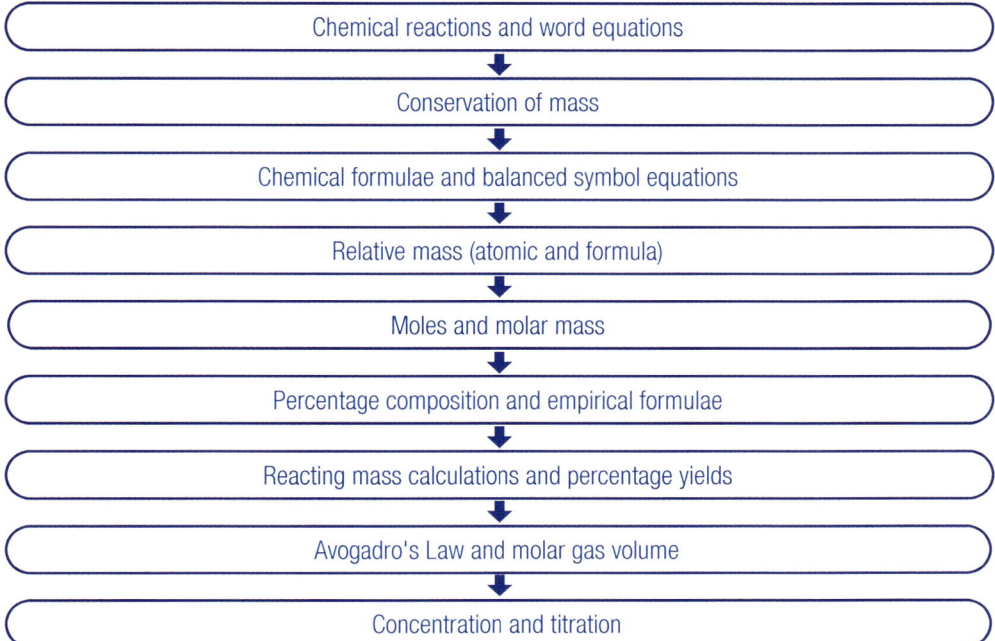

Possibly the biggest challenge that most students face is getting to grips with the idea of 'the mole' (see Chapter 3). There probably isn't one single right time to introduce the idea, but

speaking about moles, at least qualitatively, as the chemical counting unit early on can be helpful. How we measure amount of substance in chemistry, and how we account for particles in chemical reactions runs throughout the subject.

The new definition of the mole from IUPAC[1] does in some ways simplify our job:

> *The mole, symbol mol, is the SI unit of amount of substance. One mole contains exactly 6.022 1407 6 × 10^{23} elementary entities. This number is the fixed numerical value of the Avogadro constant, N_A, when expressed in the unit mol^{-1} and is called the Avogadro Number.*
>
> *The amount of substance, symbol n, of a system is a measure of the number of specified elementary entities. An elementary entity may be an atom, a molecule, an ion, an electron, any other particle or specified group of particles.*

We now have an explicit number as our 'yardstick', rather than having to jump through the conceptual hoops of elements and isotopes to get to where we need to be.

We can help students to make the conceptual link between mass and amount of substance using models and analogies. For example, consider using pots containing the same number of everyday items (buttons, wooden cubes, coins, pins, paper clips) and show how these all have different masses. Then show pots containing different substances (carbon, sulfur, copper, sodium chloride), each with a mass equivalent to the molar mass. Finally, introduce the idea of a counting unit for chemical particles (be that electrons, atoms, ions, molecules, formula units etc.) and that each pot contains the same number of particles. Students readily accept that the same number of everyday objects will have different masses, because each object has a different mass, and that we group objects by certain quantities – a pair of shoes, a dozen rolls, a century of runs. They soon pick up the same concept by analogy for the chemical substances, even when they can't see the individual particles.

Most calculations in quantitative chemistry will centre around:

$$\text{amount of substance (mol)} = \frac{\text{mass of substance(g)}}{\text{molar mass(g. mol}^{-1})}$$

Plenty of practice helps to embed this equation in students' memories, with examples and practice in all three forms of the equation ($n = m/M$, $m = n \times M$, $M = m/n$), often bringing in changes of units as well (e.g. kg instead of g).

Practical work has an important role to play in students' conceptualising of quantitative chemistry. Tasks will give them experience of chemical reactions, observations of chemical change, measurements of masses and temperatures, and sources of experimental error and measurement uncertainty. All these tell the story of how chemistry has developed over the last couple of centuries.

Several commonly used practicals can be carried out at microscale. For example:
- Determination of empirical formula – oxidation of magnesium.
- Determination of waters of crystallisation – dehydration of hydrated salts.
- Determination of enthalpy of reaction – displacement reaction between zinc and copper sulfate.

The determination of the formula of magnesium oxide by combustion of magnesium is a frequently used practical, but it can yield rather variable results. Common problems include porcelain crucibles becoming damaged and breaking during the practical, and magnesium oxide escaping when the lid is lifted. This loss of the product reduces the accuracy of the final result and makes it harder for the student to appreciate the validity of the conclusions that they are being asked to make. The microscale method outlined below actually uses a similar mass of magnesium. The key difference is the use of an inexpensive 'homemade' alternative to the expensive crucibles. The natural design of bottle top caps allows for a good flow of air with minimal loss of product.

Microscale activity 4.1: Counting by weighing

Ensure that full planning and risk assessment is carried out before attempting this activity.

The use of small inexpensive balances brings a practical introduction to the difficult concept of the mole and the Avogadro Number. In the world of paper clips, the 'Avogadro' Number is 60, or 6×10^1, hence 60 paper clips is equivalent to one 'mole' of paper clips.

Outline requirement

- small mass balance (to 2 dp)
- large weighing boat
- paper clips, 100+

Outline procedure

1. Find the mass of the weighing boat or container (M1).
2. Find the mass of the container and 6 paper clips (M2).
3. M2-M1 is the mass of 0.1 'mole' of paper clips. Calculate the mass of a 'mole' of (i.e. 60) paper clips.
4. Place all the paper clips in the weighing boat or container and find the mass (M3).
5. Calculate the mass of the paper clips (M3-M1).
6. Calculate the number of paper clips that you were provided with.
7. Arrange all your paper clips in groups of ten on a sheet of paper and find the number by counting.
8. You could do this by finding the mass of one paper clip. Does it make a difference to the result?
9. To simplify matters (depending on the group), you can zero the balance with the weighing boat on.

Chapter 4: Quantitative chemistry – moles and energy

Technical considerations – microbalances

The advent of inexpensive, robust and high-resolution mass balances over the last few years has expanded the range of techniques and practicals that can be carried out by students across the age range (Figure 4.2). For around £10 (at time of publication), a 100- or 200-gram balance, measuring accurately to 0.01 g is available from online retailers. At this price, class sets become affordable and potentially cheaper than an individual large-scale mass balance. Some additional attention is needed at the end of practical work to ensure that all devices have been safely returned. Experience from schools show that these devices remain reliable and resilient to continual use across the student age range.

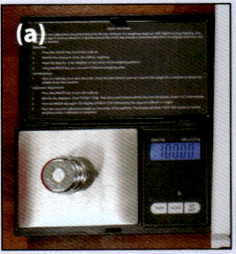 

Figure 4.2: Different mass balances available
(a) A 100 g mass on an inexpensive digital balance
(b) A 100 g mass on a traditional lab balance

Microscale activity 4.2: Determining the empirical formula of magnesium oxide

Ensure that full planning and risk assessment is carried out before attempting this activity.

This activity is based on CLEAPSS Practical Procedure PP063, *Finding the formula of magnesium oxide*[2].

Outline requirements

- magnesium ribbon (DANGER: Flammable)
- Bunsen burner
- crown bottle tops (2)
- mass balance
- eye protection
- nichrome wire
- pliers
- small pipe clay triangle
- tripod
- wire cutters

Outline method

The details of this method are contained in Figure 4.3.

Outline data processing

1. Calculate the mass of magnesium (M2-M1).
2. Calculate the mass of magnesium oxide (M3-M2).
3. Compare the experimental mass ratio of magnesium oxide to magnesium to the expected ratio (1.66).

Chapter 4: Quantitative chemistry – moles and energy

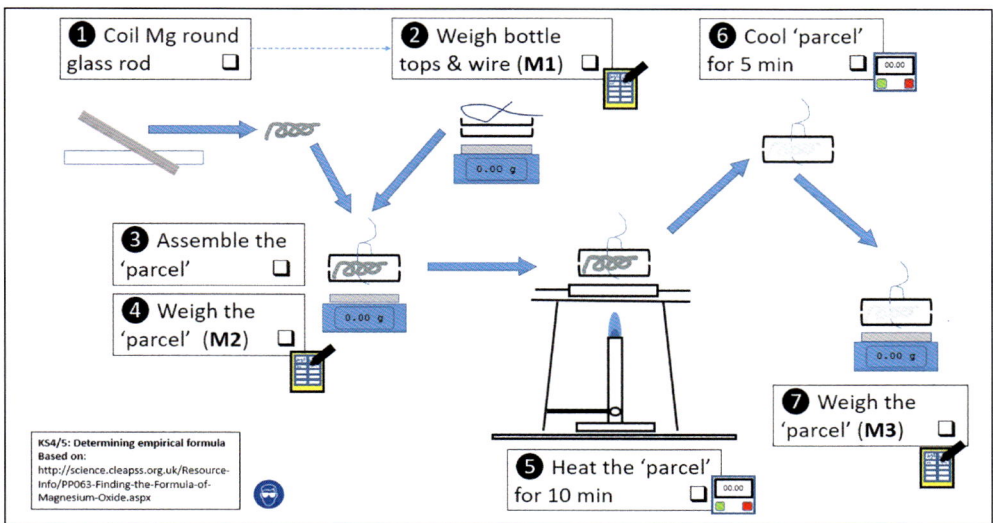

Figure 4.3: Integrated instructions for microscale activity 4.2

More advanced analysis is possible depending on where in the sequence of learning you use this activity. For example, plot a graph showing the theoretical mass ratio of magnesium oxide:magnesium for different magnesium oxide formulae, e.g. Mg_2O, MgO and MgO_2 (Figure 4.4). Alternatively, you can calculate empirical formula directly for the measured masses.

The bottle tops are a useful device for various other practicals, not least because they explicitly reduce the quantities of substances that students try to use in their experiment. This has benefits of lower costs, waste and hazards.

For example, analysis of the water of crystallisation of hydrated crystals requires pre-weighing a sample, heating to constant mass, then re-weighing and calculating the mass of anhydrous salt remaining and hence the mass of water lost.

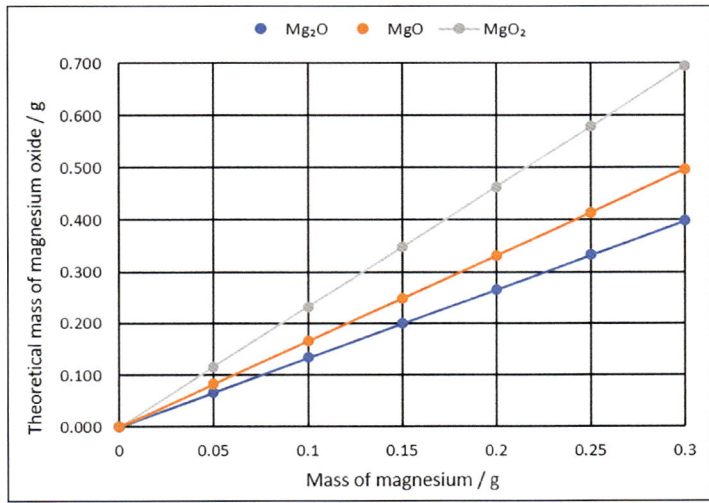

Figure 4.4: The theoretical masses of magnesium oxide produced vary depending on the empirical formula

Understanding chemistry through microscale practical work

Technical considerations – Bottle top crucibles

These 'homemade' alternatives to stainless steel or porcelain crucibles are robust (Figure 4.5). The bottle tops must be steel (not aluminium) and need to have the inner polymer removed. This is easily done with a Bunsen burner and a pair of tongs in a working fume cupboard. To make the crucible with the integral holder, drill a 6 mm hole through the flat top of the bottle top close to the edge, then thread through an M6 machine bolt and secure with a nut.

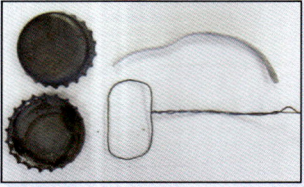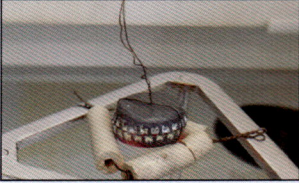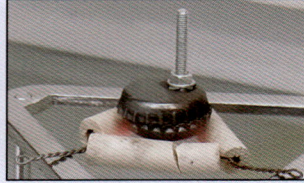

Figure 4.5: Construction and use of a bottle top crucible

Microscale activity 4.3: Determining the percentage of water in a hydrated salt

Ensure that full planning and risk assessment is carried out before attempting this activity.

This is based on CLEAPSS Practical Procedure PP039, *Finding the water in a hydrated salt*[3].

Outline requirements

- copper(II) sulfate 5-water (DANGER: Corrosive, harmful, irritant)
- bottle top crucible with bolt
- eye protection
- heatproof mat
- mass balance (to 0.01g)
- metal tongs
- spatula
- spirit burner

Outline method

1. Measure the mass of the crucible (M1).
2. Add a small amount of copper(II) sulfate 5-water to the crucible and reweigh (M2).
3. Clamp the crucible over the spirit burner.
4. Light the burner and heat the crucible until there is no further colour change.
5. Extinguish the spirit burner, allow the crucible to cool, then reweigh (M3).

Outline data processing

1. Calculate the mass of the copper(II) sulfate 5-water used = M2-M1
2. Calculate the mass of the water lost = M2-M3
3. Calculate the percentage water in the hydrated crystals = (M2-M3) × 100 / (M2 − M1)

More detailed analysis can be carried out. For example, students could calculate the empirical formula of the substance by calculating amount of substance (in mol) of the salt and water and finding the simplest whole number ratio. Other hydrated salts can be used such as iron(II) sulfate-7-water and magnesium sulfate-7-water (Epsom salts).

This method uses spirit burners rather than Bunsen burners, which limits the extent to which sulfates will decompose releasing toxic sulfur dioxide.

Technical consideration – Spirit burners

Spirit burners can be easily obtained from laboratory suppliers, but can be expensive, or too large with loose fitting tops. If tipped over, ethanol can flow out and become ignited. A cheaper alternative is the homemade version made from small-scale jam jars (Figure 4.6) – two different versions are available in the CLEAPSS GL157 guide[4]. As with the large spirit burners, do not allow the students to refill the burners and carefully demonstrate how to handle, both before and after the burner is lit.

Figure 4.6: A spirit burner made from a small jam jar

Concepts being developed 2 – energy

Energy is another tricky concept for students, and is dealt with in all three branches of school science. Recent work by the Institute of Physics highlights the difficulties:

'As there is no convenient definition of energy for beginners, the concept (of energy) needs to develop slowly until students can write about energy without making mistakes, putting the right words into the right places. It benefits from a spiral approach to teaching'[5].

Vanessa Kind summarised[6] three major difficulties that students face:
- believing that only fuels are energy 'stores';
- believing that energy can be created and used up; and
- believing that energy is released when chemical bonds break.

Understanding of chemical bonding and energetics is clearly linked, although rife with counterintuitive ideas and potential misconception. For example, the idea that reactions that release energy (exothermic) tend to have products with higher bond energies (i.e. stronger

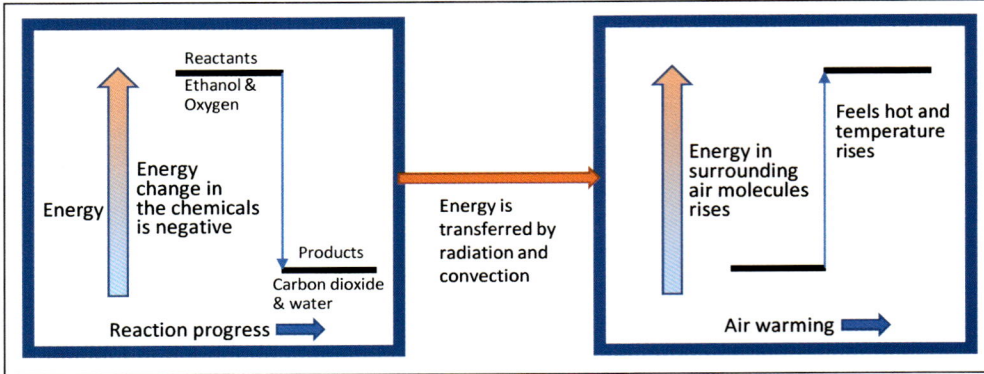

Figure 4.7: Example of energy profile diagrams for the combustion of ethanol

bonds) than the reactants is hard to grasp. We can use energy profile diagrams (Figure 4.7) to help with visualisation, and build on these to discuss catalysis and intermediates, along with many other aspects of the subject (Figure 4.8).

Practicals involving the measurement of temperature changes to allow calculation of enthalpies of reaction often involve large quantities of chemicals and significant heat loss for apparatus. These practicals are wasteful, expensive and increase the risks involved. Instead of beakers, or polystyrene cups, which have very large surface areas of the reaction volume, small glass vials wrapped in cotton wool can be used. Replacing traditional analogue (usually spirit) thermometers with digital thermometers gives better quality data. It is important for students to recognise and discuss sources of experimental error, such as heat loss. However, good experimental design from the outset should be encouraged. If practical evidence of the importance of minimising heat loss needs to be demonstrated, an enthalpy of combustion practical would be a better vehicle.

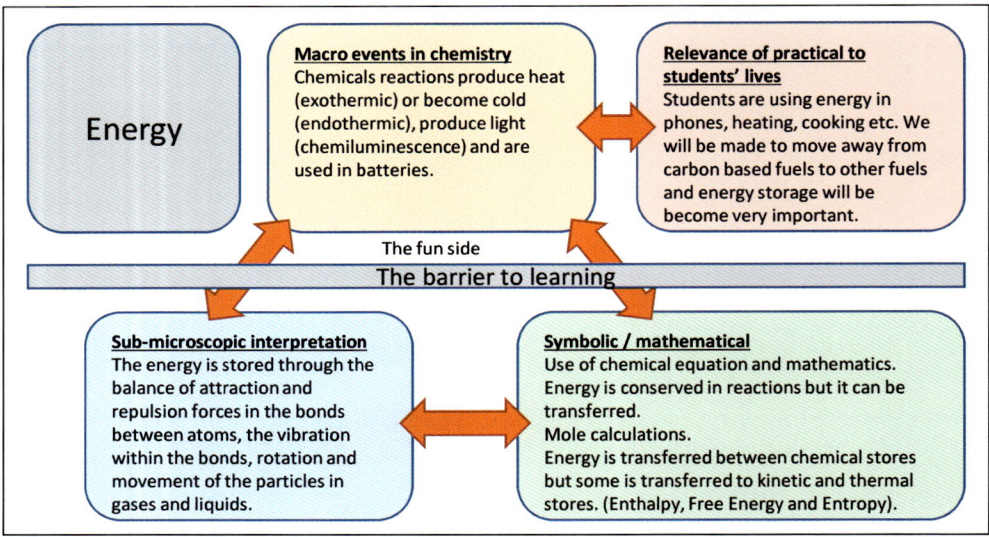

Figure 4.8: A Johnstone triangle on some aspects of energy in chemistry

Microscale activity 4.4: Determining enthalpy change of reaction

Ensure that full planning and risk assessment is carried out before attempting this activity.

Outline requirements

- 1 M copper(II) sulfate solution (DANGER: Corrosive, harmful, irritant)
- zinc powder (DANGER: flammable)
- distilled/deionised water
- goggles
- cotton wool
- digital thermometer
- small glass vials (~14 cm^3)
- 10 cm^3 measuring cylinder
- 100 cm^3 plastic beaker

Outline method

1. Set up the reaction container with a glass vial surrounded by cotton wool in a 100 cm^3 beaker (Figure 4.9).
2. Add 10 cm^3 of copper(II) sulfate solution to the vial and measure the temperature.
3. Add a fixed mass of zinc powder to the vial, stir and measure the maximum temperature.
4. Waste should be collected and disposed of carefully. Repeat the practical with different concentrations of the copper sulfate solution.

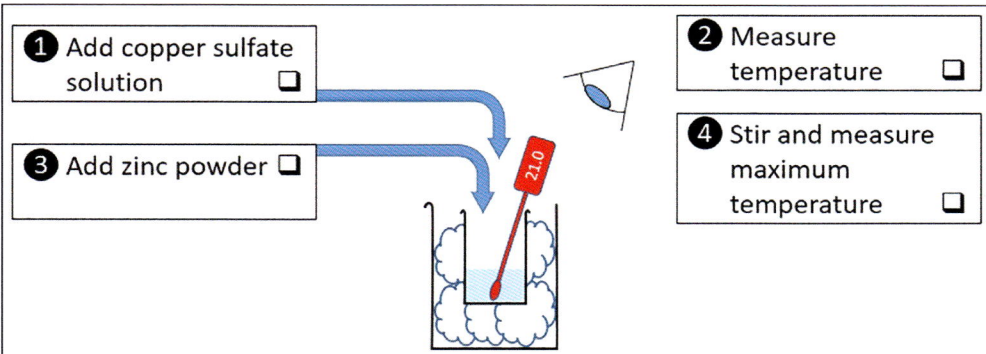

Figure 4.9: Integrated instructions for microscale activity 4.3

Outline data analysis

For 10 cm^3 samples of copper sulfate of different concentration and excess zinc, the graph shows the relationship between concentration and temperature change (Figure 4.10). Enthalpy change for the reaction (ΔH) can be determined from the equation

$$\Delta H = -(\textit{specific heat capacity}) \times 1000 \times \textit{gradient}$$
$$= -4.18 \times 1000 \times 33.75 = -141 \ kJmol^{-1}$$

This method substantially reduces the costs and disposal issues of this practical being carried out as a full class practical, especially if multiple repeats and multiple classes across a year group are considered. If time is limited, then class data could be pooled in a spreadsheet

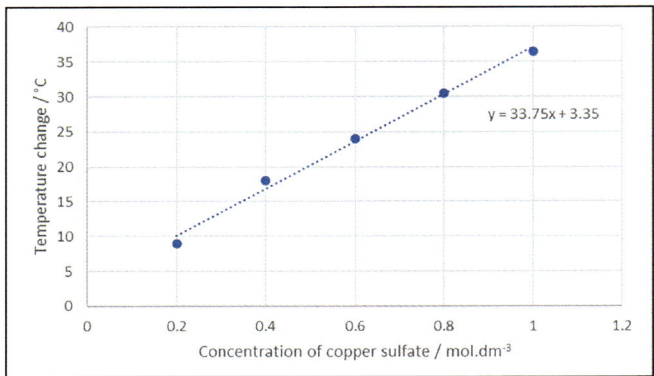

Figure 4.10: Sample data and analysis for microscale activity 4.4

and one graph drawn from mean averages of each volume point. As with other microscale alternatives to traditional practical activities, students get to the data much more quickly, allowing you to spend more time supporting the students' development of analysis skills and discussing the chemistry involved.

Technical consideration – Digital thermometers

Digital thermometers are now widely available through online suppliers for a few pounds, and measure accurately to 0.1 °C (Figure 4.11). Depending on the quality of the thermometer, the response time (how long it takes the reading to reach the measured temperature) can be quicker than analogue (spirit) thermometers. The thermometers can be top heavy, so equipment may need to be clamped. Spirit thermometers may be very inaccurate. However, since students are reading a difference in temperature rather than taking an absolute temperature, this may not be an issue. To compare thermometers, place some in ice water and some in very hot water and compare the measured values.

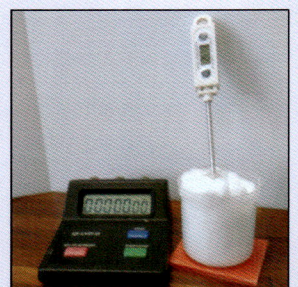 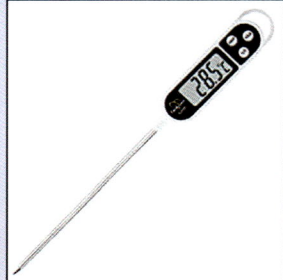

Figure 4.11: Digital thermometers

Teacher activities

Consider other areas of the curriculum where students are measuring amounts of substance and temperature changes and where microscale equipment and amounts could be used. For example:

- Determining the enthalpy of neutralisation – reaction of 2 M sodium hydroxide and 2 M hydrochloric acid; repeat with 2 M ethanoic acid in place of hydrochloric acid; carry out a thermometric titration to provide evidence of mole ratios of acid to alkali.
- Determining the percentage yield of a reaction – synthesis of aspirin or preparation of hydrated copper sulfate crystals from a known mass of copper oxide.
- Determining the enthalpy change of thermal decomposition of potassium hydrogen carbonate – calculation of this enthalpy change by the indirect route by reaction of potassium hydrogen carbonate and potassium carbonate with acid.
- Determining the reduction in mass on heating sodium hydrogen carbonate; establish whether the final product is sodium hydroxide or sodium carbonate.
- Determining enthalpies of combustion of different alcohols – consider whether small aluminium drinks cans (150 ml mixers) can be used, and whether larger volumes of water and smaller temperature rises give more accurate results.

In each case, consider whether there are true advantages to carrying out the activities on the larger scale, and whether the extra expense, hazard and disposal issues are justified. Look at the expected learning outcomes of the practicals. Are they self-demonstration practicals where students are gaining experience with handling particular reactions or substances? Is the quality of the data of most importance and hence larger amounts may be justified? Or are the students looking for trends in data that will clearly show up using the microscale method?

Further reading

Aspirin – A curriculum resource for post-16 chemistry and science courses, RSC. https://edu.rsc.org/resources/aspirin-book/56.article

References

[1] https://goldbook.iupac.org/terms/view/M03980 (Accessed August 2021)

[2] http://science.cleapss.org.uk/Resource-Info/PP063-Finding-the-Formula-of-Magnesium-Oxide.aspx [Membership required] (Accessed August 2021)

[3] http://science.cleapss.org.uk/Resource-Info/PP039-Finding-the-water-in-a-hydrated-salt.aspx [Membership required] (Accessed August 2021)

[4] http://science.cleapss.org.uk/Resource-Info/GL157-Make-it-guide-a-spirit-burner.aspx [Membership required] (Accessed August 2021)

[5] https://spark.iop.org/words-used-describe-energy (Accessed August 2021)

[6] *Beyond Appearances: Students' misconceptions about basic chemical ideas, 2nd Edition*; https://edu.rsc.org/resources/beyond-appearances/2202.article (Accessed August 2021)

Chapter 5: Titration

'Microscale chemistry is an excellent resource for chemistry teachers. It provides straightforward guidance and easy to use equipment [for use] in a classroom. The microscale approach opens up the chance for pupils studying chemistry at all levels to carry out whole class practical work where resources are in short supply or where technical support might be limited. Furthermore, by using smaller volume of reagents, potential hazards are reduced, which means it is safer for everybody and the volume of waste to dispose of is minimal. For example, the small-scale version of the diffusion and reaction between lead nitrate and potassium iodide can be observed really clearly because there is minimum kit getting in the way of the point of interest. This reaction close up never fails to impress'

(Dr. Elaine Wilson, Associate Professor, Faculty of Education, University of Cambridge).

Concepts being developed

Learning about titration is a great example of where microscale chemistry can be used to complement, rather than replace, the traditional practical technique. Having said that, readers who try this approach may be surprised at the results when compared to the traditional method. The gravimetric titration[1] approach using the mass balance can be controversial, since it is very different from the traditional procedure used. However, when volumetric titration was introduced in the 1850s, precision balances had not been invented.[2]

Titration is an analytical technique that allows you to determine the concentration of a solution. This is achieved by adding a titrating solution (the titrant) of known concentration to a sample of the solution being analysed (the analyte). The titrant is added until there is an observable or measurable change to the mixture. Often, an indicator is used in the mixture so that a colour change can be seen. In some cases, the reaction mixture is self-indicating such as with many redox titrations and, in other cases, measurements such as temperature or electrical conductivity can be used.

Chapter 5: Titration

Many students will be introduced to titration with acid-base titrations. The simplest to carry out is the titration of sodium hydroxide with hydrochloric acid, using either phenolphthalein or methyl orange as an indicator.

Consider the following narrative of how we may explain to students how to carry out a titration:

Wash the volumetric pipette with 0.1 M sodium hydroxide solution and the burette with the 0.1 M hydrochloric acid solution. Add a 25 cm³ sample of the sodium hydroxide to a 250 cm³ conical flask using a volumetric pipette, along with a few drops of phenolphthalein indicator solution, then swirl the flask to mix. Fill a burette with the hydrochloric acid solution using a funnel, ensuring that the tap is closed and you are able to safely transfer the liquids below eye level. Remove the funnel, and arrange the burette over a waste beaker using a burette clamp. Run the acid into the waste beaker to remove any air from the burette nozzle. Take an initial burette reading to an accuracy of 0.05 cm³. Carry out a rough titration by adding the acid to the conical flask with swirling, until you see the colour change, noting the volume of acid you have added. Rinse out the conical flask with tap, then distilled, water between titrations. Now carry out multiple accurate titrations, ensuring that acid is added dropwise towards the end point, until you have at least two with concordant results (within 0.1 cm³).

Within this last paragraph we can see why this technique can be so problematic for students to learn and master. We are asking them to think about, amongst other things:

- how indicators work;
- how to measure accurately with a volumetric pipette;
- how to set up delicate glassware;
- how to safely transfer liquids;
- simultaneously carrying out three operations – controlling the tap, swirling the flask and observing the solution;
- making accurate readings on the burette; and
- accurately determining the end point based on a colour change (then reproducing it).

All of this is a recipe for an overloaded working memory and a student who is unlikely to be effectively carrying out the technique, let alone productively learning what we want them to[3]. To help students, we can break the overall titration task down into smaller parts, and give them time to understand and practise each part, before moving on to the next.

An alternative and more useful sequence would be:

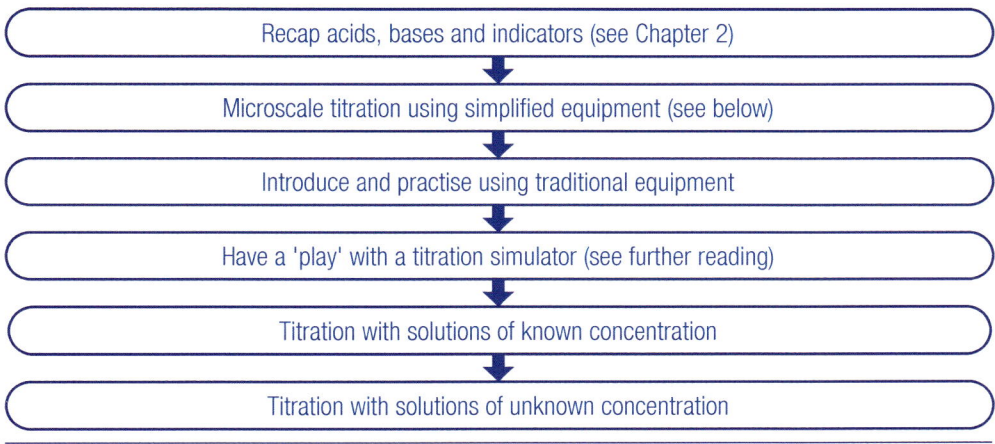

Understanding chemistry through microscale practical work

47

Chapter 5: Titration

There are two useful microscale titration techniques, each with different learning objectives:
- Gravimetric titration – using a mass balance to weigh the reaction vessel before and after the titration, and using these data to calculate the concentration of the analyte.
- Syringe/pipette titration – using a syringe to add the titrant to the reaction vessel so that volumes can be measured directly.

The choice of which technique to use, or in which order to use them, will depend on when you are choosing to introduce the data analysis. If you are starting with a theoretical approach first, so that students know how to process the volume data to calculate concentrations, then the syringe/pipette titration will work well. If you are starting with the practical approach first, so that students understand where the data come from, the gravimetric titration works well. This is very much a personal choice, since learning titration will usually take several lessons, working towards the same end point.

As well as the advantages of breaking down a complex task into simpler tasks, these microscale activities give students an opportunity to work on their fine motor skills, which are important when using the larger scale equipment. Learning to add titrant solution carefully, either by the turn of the clamp bolt, or the gentle pressing of the syringe plunger, can help with manipulation of the burette tap.

Microscale activity 5.1: The mass of a drop of liquid

Ensure that full planning and risk assessment is carried out before attempting this activity.

Outline requirements

- ethanol (DANGER: Highly flammable)
- water
- boss and clamp
- eye protection
- different type of plastic pipettes
- plastic Petri dish (small)
- retort stand
- small mass balance (to 0.01 g)

Outline method

1. Record the mass of the Petri dish.
2. Fill the plastic pipette with water and lightly clamp it.
3. Place the Petri dish under the clamped pipette.
4. Add 10 drops of water (Note that, if by mistake 9 or 11 drops are added, it does not matter because the value in the table is simply altered).
5. Record the mass of the Petri dish and 10 drops of water.
6. Repeat this procedure another 6 to 10 times so that many more drops have been added. The mass is recorded for each group of 10 drops that have been added.
7. Draw a graph with number of drops on the x axis and mass (g) on the y axis. The gradient of this graph will provide the average mass of one drop.
8. The procedure can be carried out with ethanol.

Chapter 5: Titration

Microscale activity 5.2: Gravimetric titration

Ensure that full planning and risk assessment is carried out before attempting this activity.

This is based on CLEAPSS Practical procedure PP019, *Analysis of vinegar by small-scale titration*[4].

Outline requirements

- 0.25-0.4 M sodium hydroxide (WARNING: Irritant)
- 0.4 M hydrochloric acid
- 0.1% phenolphthalein or 0.1% methyl orange solution (WARNING/DANGER: (Highly flammable)
- eye protection
- plastic pipettes (preferably one with a fine tip) (2)
- retort stand, boss and clamp
- small glass vial (see Chapter 6)
- small mass balance (to 0.01 g)

Outline method

1. Add one drop of indicator to the glass vial, weigh and note the mass (M1).
2. Add about 1 cm^3 hydrochloric acid to the vial, weigh and note the mass (M2).
3. Fill the plastic pipette with sodium hydroxide solution and lightly clamp.
4. Place the vial under the clamped pipette (Figure 5.1).
5. Add sodium hydroxide, drop by drop, to the vial by slowly turning the boss nut and swirl the flask to mix.
6. Continue until the end point (indicated by the appropriate colour change depending on the indicator used).
7. Weigh the vial and note the mass (M3).

Figure 5.1: The small-scale gravimetric titration set-up

Outline data processing

Assume that all solutions have the same density as water.

The volume of the solution added (in cm^3) will be equivalent to the mass of solution (in g).

$$m(HCl) = M2 - M1; \quad m(NaOH) = M3 - M2$$
$$\text{So } c(NaOH) = 0.4 \times ((M2 - M1) / (M3 - M2))$$

We can present this formula to students as a 'black box' at this stage, one into which they put their mass values, to get the concentration of the sodium hydroxide. Since the idea that they are developing is about collecting quantitative data to allow calculation of a property, the 'nuts

Understanding chemistry through microscale practical work

Chapter 5: Titration

and bolts' of the calculation doesn't need to be the focus. Even better, set up a spreadsheet into which they can type their values and which will automatically calculate the concentration. This gives the advantage of allowing a discussion of the range of data collected across the class.

This activity can also be presented as a context-based investigation, by changing the hydrochloric acid for diluted vinegar. Different samples can be presented as though from chip shops around the town, one of which is diluting its stock to increase its profits! The CLEAPSS *Practical procedure PP020, Analysis of vinegar by titration*[5], is a useful source of further information.

Microscale activity 5.3: Titration using a syringe-pipette

Ensure that full planning and risk assessment is carried out before attempting this activity.

Outline requirements

- 0.4 M hydrochloric acid
- 0.1% phenolphthalein or 0.1% methyl orange solution (WARNING/ DANGER: (Highly) flammable)
- 0.25-0.4 M sodium hydroxide (WARNING: Irritant)

- eye protection
- boss and clamp
- 2 cm^3 graduated glass pipette connected to a 10 cm^3 syringe by silicone tubing
- small glass vial
- retort stand
- 1 cm^3 syringe

Outline method

1. Add 1 cm^3 of sodium hydroxide solution into the vial using the 1 cm^3 syringe.
2. Add one drop of indicator solution to the vial – agitate to mix.
3. Draw hydrochloric acid into the pipette using the syringe.
4. Gently clamp the pipette and place the vial underneath the tip.
5. Measure and record the initial volume in the pipette.
6. Gently depress the plunger of the 10 cm^3 syringe to add hydrochloric acid dropwise to the vial, swirling the mixture after each addition, until the end point is reached.
7. Record the final volume in the pipette.

This method uses equipment that the students are likely to be more familiar with. Syringes are commonly supplied, for example, with children's painkiller syrups. Direct volume measurements of both the sodium hydroxide, and the hydrochloric acid by volume difference, allows students to carry out titration calculations directly. A further simplification can be made to the equipment by replacing the graduated pipette with a microlitre pipette tip and a smaller syringe. The volume readings will not be to such a high resolution, but readings may be easier.

Outline data analysis

Volumetric (or gravimetric) analysis is a useful place in a chemistry course to discuss concepts around the quality of data. Titration data are usually collected until at least two concordant

Technical considerations – Microburettes

Microburettes can be assembled with a graduated 2 cm^3 pipette, a short length of silicone tubing and a 5 cm^3 syringe. Connect the pipette and syringe together with the tubing and clamp the pipette vertically. Draw the titrant into the pipette by lowering the pipette tip into the stock solution and slowly drawing back the plunger on the syringe. Students will need to practice dispensing the titrant, drop by drop, using the syringe. Holding the body of the syringe with both hands and pushing against the plunger and the opposing thumb simultaneously can help with the control.

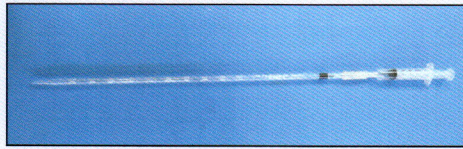

Figure 5.2: A microburette

results have been achieved. For large-scale titrations, this would be expected to be two titre values within 0.10 cm^3, with a mean average titre of concordant results then calculated.

The percentage uncertainly of experiment values is calculated as:

$$\text{percentage uncertainty} = \frac{\text{measurement uncertainty}}{\text{measurement}} \times 100\%$$

As such, experimentally, we look to decrease the measurement uncertainty and increase the measurement values to minimise the percentage uncertainty. The two microscale techniques yield percentage uncertainties of approximately 2% (for titres of approximately 1 cm^3) compared with approximately 0.4% for large-scale burettes (for titres of approximately 25 cm^3). This higher uncertainty can then be reduced by increasing the volumes of the titrants used, and by using equipment with higher resolution, with the inevitable increase in cost. This trade-off between cost of analysis and accuracy/precision of data is a useful discussion point for students. It is worth noting that microscale titrations are quicker than using burettes, so more repeats can be completed in the same time.

Technical considerations – measuring devices

Using different volumetric measuring devices in titration will give the student an appreciation of the range of apparatus available for making measurements. They will have been exposed early in their chemistry education to the difference between inaccurate gross measurements in beakers, to the more accurate measuring cylinders. By using graduated pipettes, they will appreciate that even small volumes can be accurately measured, and to a high resolution. This concept can be taken further by discussion, looking at how microlitres and even nanolitres of solutions can be analysed by techniques such as crystallography and mass spectrometry.

Teacher activities

Consider all the other areas where we use titration in teaching chemistry and how microscale titration could be used. Learning to use a burette and volumetric pipette are valuable skills, but do we necessarily need to use them when carrying out all the volumetric analysis in a course? During a post-16 course, a student may:

- determine acid or alkali concentration by acid-base titration
- determine oxidant or reductant concentration by redox titration
- determine cation concentration by chelation titration
- determine equilibrium constants.

In each of these cases, we need to consider what is the key learning outcome. Is it the ability of the student to collect accurate and reliable data, or is it the underlying chemistry of how the reaction is occurring and the associated data analysis and interpretation? Could data be collected more quickly, and cheaply, by doing some of these analyses on a microscale?

Consider whether the accuracy of the data gained by volumetric analysis using 50 cm^3 burettes is better than gravimetric analysis. Provide students with the equipment to carry out both types of analysis and compare the results. Ask the student to consider the quality of the data by calculating percentage uncertainties for their data.

Further reading

Near as makes no difference: https://www.sserc.org.uk/wp-content/uploads/2020/08/Bulletin-257-complete.pdf (Accessed August 2021)

Paterson, D. (2021) 'Making learning child's play', *Education in Chemistry*, March, https://edu.rsc.org/feature/make-learning-chemistry-childs-play/4013361.article (Accessed August 2021)

References

[1] https://uwaterloo.ca/chem13-news-magazine/december-2012-january-2013/activities/gravimetric-titration-simple-buret-free-and-high-precision (Accessed August 2021)

[2] https://www.chemistryworld.com/opinion/classic-kit-mohrs-burette/3004927.article (Accessed August 2021)

[3] See *Making practical work more effective*, David Paterson, RSC Education in Chemistry, September 2017, https://edu.rsc.org/feature/making-practical-work-more-effective/3008027.article (Accessed October 2021)

[4] http://science.cleapss.org.uk/Resource-Info/PP019-Analysis-of-vinegar-small-scale.aspx (Accessed August 2021) An open-access video is available at http://youtube.com/watch?v=YzipDbdzgTc (Accessed August 2021) [Membership required]

[5] http://science.cleapss.org.uk/Resource-Info/PP020-Analysis-of-vinegar-by-titration.aspx [Membership required] (Accessed August 2021)

Chapter 6: Rates of reaction and dynamic equilibrium

'It's good to do the microscale experiment yourself. It revealed my own misconceptions and let me see the experiment through the students' eyes!'

'Microscale practicals provide a unique opportunity to develop a range of working scientifically skills that form an integral part of the national curriculum'

'As the practicals are often quicker and give the expected result, they can cement subject knowledge rather than embed misconceptions'

(Student teachers at St Mary's University, Twickenham, London, UK).

Concepts being developed

Rates of reaction, or reaction kinetics, bring together ideas about how we can measure how quickly reactions are proceeding, with ideas about how particles are interacting at a nanoscopic level. The central theory is 'collision theory', which identifies the factors required for reaction (collision of particles, sufficient energy of collision, correct orientation of particles), and the factors that affect the rate of reaction (concentration, pressure, surface area, light intensity, temperature, presence of catalysts) (Figure 6.1).

We can represent reactions with symbolic equations and relate this directly to experimental data with tables of data, graphic analysis and rate equations. Students can struggle to make links between these multiple sources of information, and to see the connection between gas collected in inverted measuring cylinders and the graphs and equations they are writing. Careful sequencing of ideas and activities will help students develop their conceptual framework over time.

Chapter 6: Rates of reaction and dynamic equilibrium

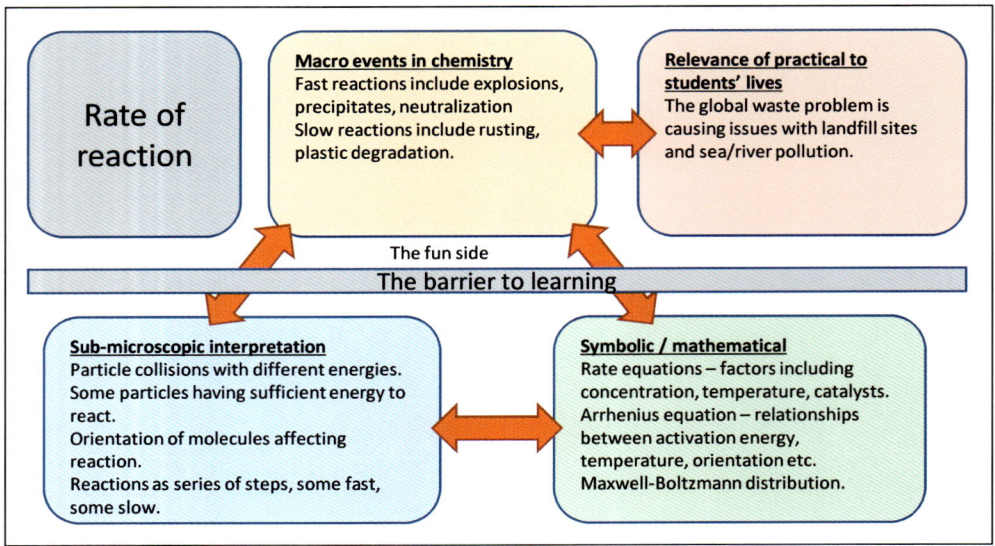

Figure 6.1: A Johnstone triangle on rates of reaction

A possible teaching sequence through the rates of reaction topic will generally integrate several contexts, such as cooking, explosions, production of industrial chemicals (such as ammonia and sulfuric acid), and biochemical reactions. A common sequence would be:

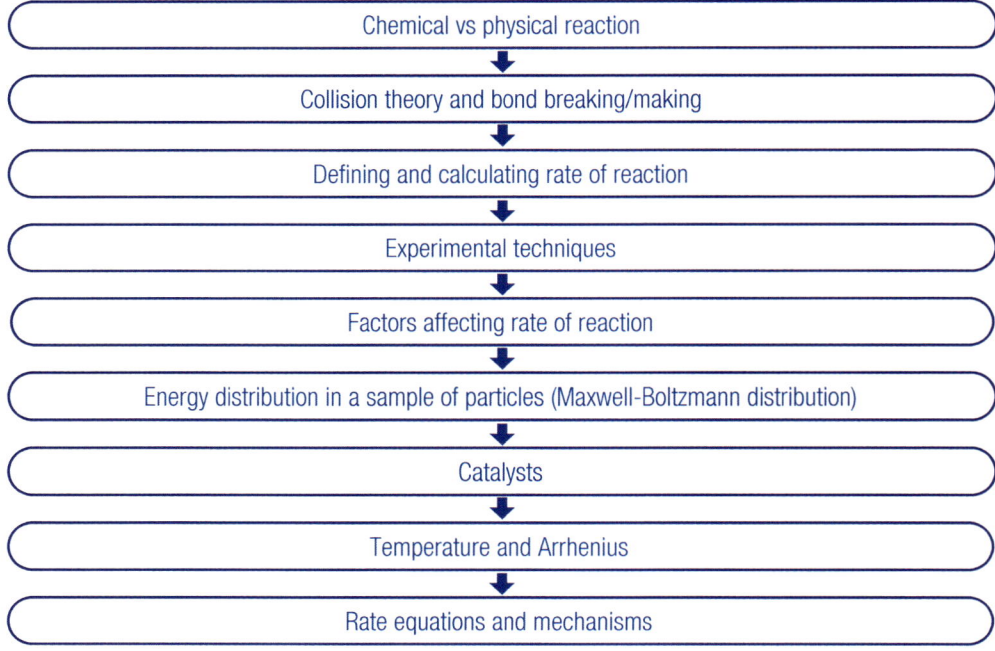

Chapter 6: Rates of reaction and dynamic equilibrium

Alongside conceptual development, students should be developing their manipulative and experimental skills – the craft side of learning science. Experimental data may be collected as:
- mass using a top pan balance;
- capture of gas produced using a gas syringe (or plastic syringe with a three-way tap) or over water in a measuring cylinder or burette; and/or
- production of a coloured substance or precipitate.

Simple reactions can be used to learn the processes and skills of data collection, and to introduce a variety of methods. This can help counteract the impetus towards only teaching the 'exam board approved' methods. Common reactions include the reaction of calcium carbonate or magnesium with hydrochloric acid. Large-scale 'traditional' methods remain useful, as either hands-on or teacher demonstration, since these methods are commonly seen in exam questions. Modifying these reactions to make them microscale would have limited efficacy, since the uncertainties around data collected would increase significantly.

One advantage of carrying out some rates experiments on a microscale is in limiting the production of hazardous products (see activities 6.1 and 6.2 below). For example, the reaction between acid and thiosulfate is commonly used as a rates investigation. Insoluble sulfur forms in the reaction mixture, masking a cross below the reaction vessel, as the mixture becomes opaque. Rate is calculated as the inverse of time (1/t). A co-product in the reaction is the toxic gas sulfur dioxide. Carrying out the reaction at a larger scale presents significant risks to staff and students alike, especially at higher temperatures. Other reactions can be used for initial rates investigations, including the iodine clock reaction, such as the Landolt and the Harcourt-Essen methods. Carried out on a large scale, the reaction mixtures can prove hazardous and require careful treatment.

Knowledge of rates of reaction also improves the energy diagram developed in Chapter 4 (Figure 6.2). A ball at the top of a hill has potential energy because of its position, and will spontaneously roll downhill. If a small piece of wood is placed in front of the ball, it will remain in position until a person pushes the ball over the wood and then it rolls down the hill to a position of lower potential energy. Mixtures of hydrogen and oxygen, or the highly flammable ethanol in air, can be stored at room temperature forever. Apply just a little energy in the form of a spark and there will be an explosion – very fast reactions with low activation energies. The rapid reactions of adding an acid to an alkali at room temperature have hardly any activation energy at all. The more the diagrams are repeated and improved, the better the students' understanding of chemical reactions.

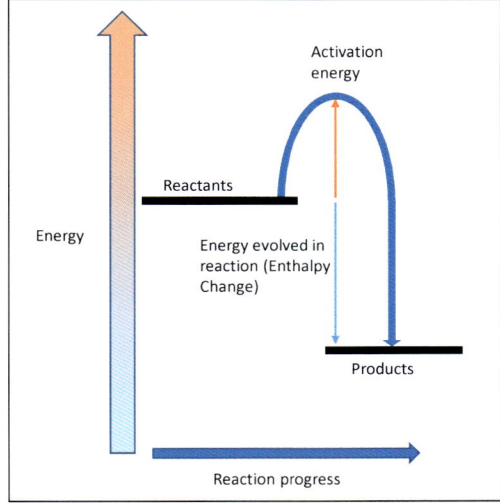

Figure 6.2: A reaction progress graph for an exothermic reaction

Understanding chemistry through microscale practical work

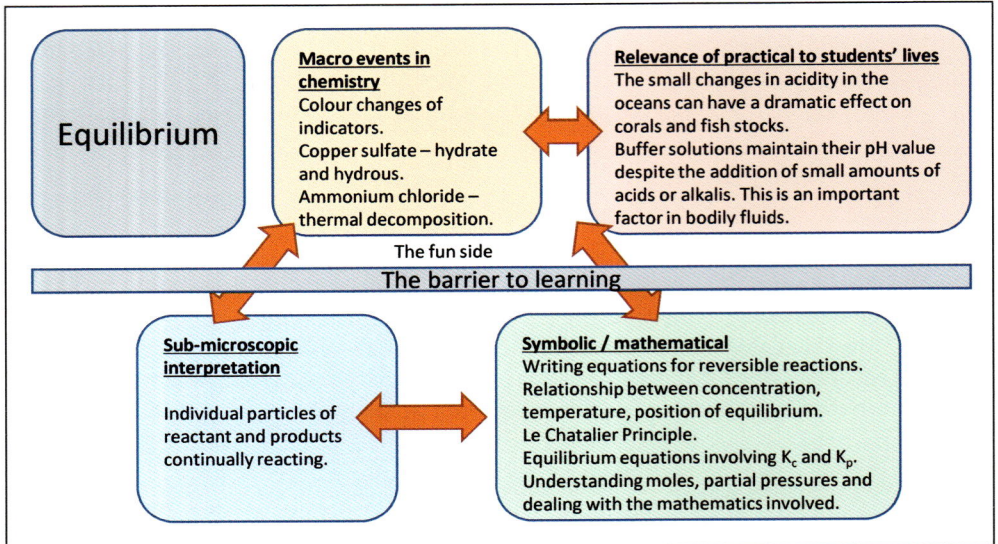

Figure 6.3: A Johnstone triangle on equilibrium

Equilibrium

Once students have a firm understanding of how reactions are measured and analysed, the ideas of chemical equilibrium can be introduced and developed (see activities 6.3 and 6.4 below) (Figure 6.3). Chemical equilibrium looks at how reversible reactions behave in a closed system. Dynamic equilibrium is reached when both forward and reverse reactions are happening at the same rate. There is no change at the macroscopic level – concentrations of reactants and product(s) remain steady, although not always equal. At the nanoscopic level, particles are continually reacting, forming reactants and products. This idea of no observable change masking the continual change at the particle level is very challenging for many students. One particular hindrance is that students are exposed to equilibrium in physics quite early on, and tend to picture children sitting on seesaws when we discuss chemical equilibrium. These misconceptions should be highlighted and discussed clearly.

Reactions carried out on the microscale tend to be quicker with clearer observations, leaving more lesson time for students to consider and discuss these observations, and to develop their chemical concepts. The energy diagram (Figure 6.4) helps to understand the difficult concept of equilibrium where collisions between particles are resulting in both forward and backward reactions but the concentrations of the 'reactants' and 'products' are not altering. Usually, the difference in chemical energy between the 'reactants' and 'products' and the activation energies of the forward and backward reactions is small.

Chapter 6: Rates of reaction and dynamic equilibrium

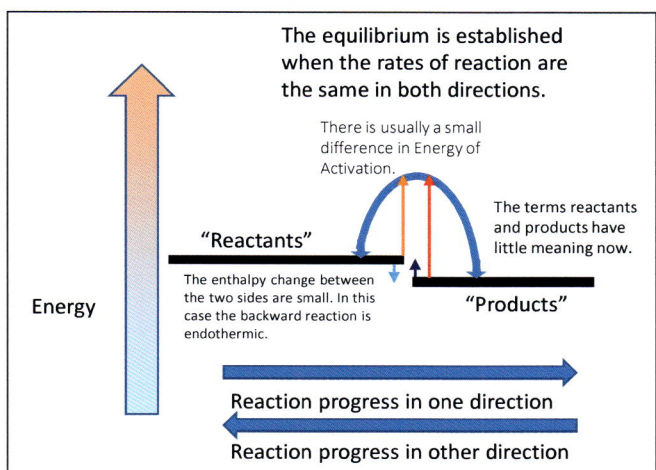

Figure 6.4: A reaction profile diagram for a reversible reaction

A common teaching sequence would be:

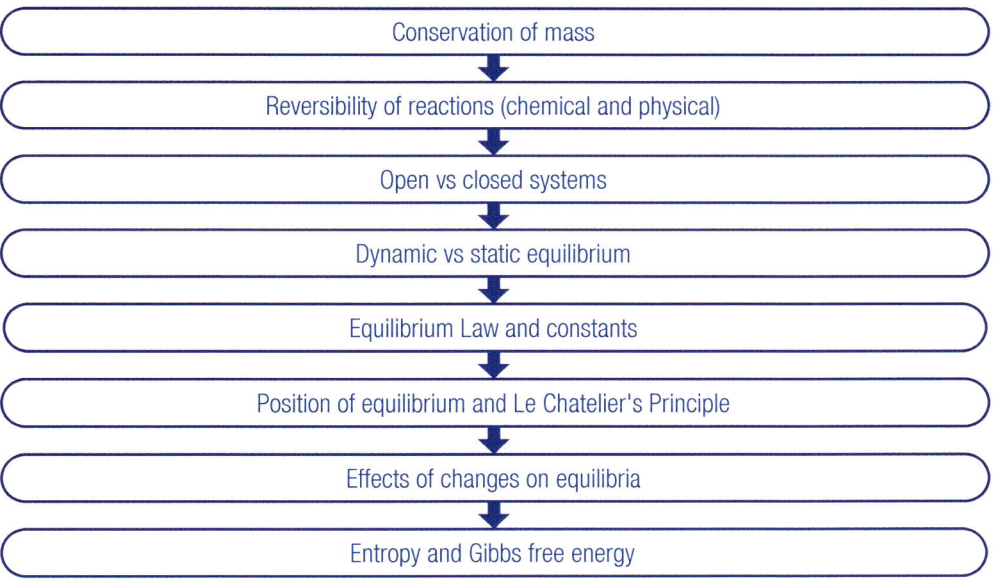

Understanding chemistry through microscale practical work

Chapter 6: Rates of reaction and dynamic equilibrium

Microscale activity 6.1: Investigating the effect of concentration on rate of reaction

Ensure that full planning and risk assessment is carried out before attempting this activity.

This is based on CLEAPSS Practical Procedure PP041, *The thiosulfate-acid reaction: rate and concentration*[1].

Outline requirements

- distilled/deionised water in wash bottle
- 1 M hydrochloric acid
- phenolphthalein solution (DANGER: Highly flammable)
- 0.1 M sodium thiosulfate solution
- 0.5 M sodium carbonate solution
- eye protection
- glass marker pen
- large beaker
- measuring cylinder
- plastic dropping pipette
- small glass vials (2)
- timer
- vial box

Outline method

1. Make the reaction stop bath with sodium carbonate solution and phenolphthalein.
2. Draw a cross on the base of one glass vial – this is the reaction vial.
3. Fill the other vial with hydrochloric acid – this is the acid vial.
4. Transfer 1 cm^3 of hydrochloric acid from the acid vial to the reaction vial.
5. Make up a 10 cm^3 solution of sodium thiosulfate of the required concentration in the measuring cylinder using the stock sodium thiosulfate solution and distilled water.
6. Pour the sodium thiosulfate solution into the reaction vial and simultaneously start the timer.
7. Stop the timer when the cross is no longer visible.
8. Pour the reaction mixture into the stop bath, clean the vial and repeat with different concentrations of sodium thiosulfate solution.

Outline data analysis

- Calculate the rate of each reaction using the equation: rate = 1 / time.
- Plot a graph of concentration of sodium thiosulfate solution used against rate of reaction.
- Analyse the graph and identify the relationship between concentration of thiosulfate and rate as directly proportional, hence the reaction is first order with respect to thiosulfate.

This method has several practical and conceptual advantages over the commonly-used large-scale reaction.

Practical: The total amount of each reagent used is significantly lower, reducing the materials cost and time taken for preparation. Since the volumes are reduced, the total amount of toxic sulfur dioxide produced is significantly reduced, lowering the risk of adverse effects on students and teachers. By including a stop bath (that stops the reaction by neutralising the hydrochloric acid and neutralises the acidic sulfur dioxide), further release of sulfur dioxide is prevented. The amount of sulfur, which can be difficult to clean from glassware, is also reduced. The risk of cross-contamination of the reagents is also significantly reduced since students are measuring the different solutions with difference devices.

Conceptual: The students will be making similar observations to the larger-scale version, i.e. watching through the solution until the cross is no longer visible. However, the time taken to collect a full set of results is significantly reduced. This gives students the opportunity to repeat the investigation, reducing the effect of random error and anomalous results. Data processing and analysis is also possible within one practical session, since data collection time is reduced. Students will benefit from a clearer understanding of the link between data and analysis if the gap between these two activities is reduced.

This activity can easily be extended to investigate the effect of temperature on the rate of reaction. By adding hot or cold water to the vial box, the reaction mixture can be heated or cooled quickly. Data in the CLEAPSS *Practical Procedure PP141*[2] (Figure 6.5) shows that the activation energy of this reaction can be determined using the Arrhenius equations.

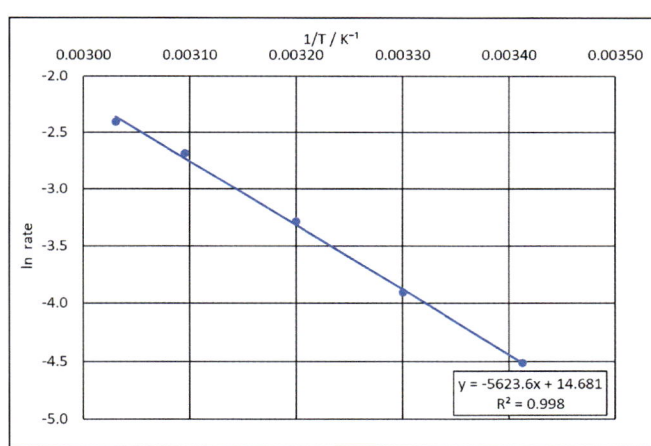

Figure 6.5: An Arrhenius plot of data collected from the acid-thiosulfate reaction at different temperatures

Technical consideration – Small reaction vials

Many reactions carried out on a smaller scale are best done in flat-bottomed glass vials (Figure 6.6). These are usually sold as 'rolled rim vials', which allow for attachment of a snap lid. These vials cost about the same as test tubes and do not require test tube racks.

Figure 6.6: Small glass vials for microscale reactions

Chapter 6: Rates of reaction and dynamic equilibrium

Technical consideration – Reaction boxes

If racks are required for the vials, these can be simply made from plastic food boxes (Figure 6.7). Holes need to be made in the lid of the boxes to allow the vials to fit through. These holes can be cut with craft knives, or melted through with a large diameter cork-borer or heated coin. Work in a fume cupboard, hold a large coin in a roaring blue flame with tongs for 20 seconds, then carefully melt a hole in the lids. Check the fit and expand the hole as necessary to allow the vial to fit through snugly without requiring excessive force.

As well as making useful racks for the vials, hot or cold water can be poured into the container itself, making water baths for heating or cooling reaction mixtures. Small holes can also be made for a thermometer to monitor the temperature of the water.

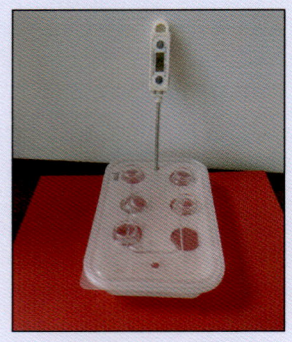

Figure 6.7: Reaction boxes used for microscale practicals

Microscale activity 6.2: An iodine clock reaction

Ensure that full planning and risk assessment is carried out before attempting this activity.

Outline requirements

- solutions in dropper bottles
  - distilled/deionised water
  - 0.025 M potassium iodate(V)
  - 0.4 M sodium hydroxide (WARNING: Irritant)
  - 0.025 M sodium metabisulfite
  - 1% starch
- eye protection
- spotting tile
- timer

Outline method

1. Add 1 drop of starch solution to each of four wells.
2. Add 8 drops of potassium iodate(V) to each of the four wells.
3. Add 2, 4 and 6 drops of water to the second, third and fourth well respectively.
4. Add 8 drops of sodium metabisulfite to the first well and immediately start the timer.
5. Stop the timer when the solution turns blue/black. Record the time.
6. Add a few drops of sodium hydroxide to the blue/black solution to prevent iodine staining.
7. Repeat steps 4-6 with 6, 4 and 2 drops of sodium metabisulfite in the second, third and fourth wells respectively.

Outline data analysis

- Calculate rate of reaction using the equation: rate = 1 / time.
- Plot a graph of number of drops of sodium metabisulfite vs rate of reaction.
- Draw a line of best fit.
- Interpret your data/graph to deduce the order of reaction with respect to sodium metabisulfite.

Extend the practical by modifying the concentration of potassium iodate(V) while leaving the concentration of sodium metabisulfite unchanged. The advantages of carrying out a clock reaction on the microscale compared with larger scale are similar to those discussed for Activity 6.1.

This activity is worth testing before use with students. Stocks of sodium metabisulfite do not store well, and should be made fresh.

Microscale activity 6.3: Demonstration of conservation of mass

Ensure that full planning and risk assessment is carried out before attempting this activity.

Further information can be found in the RSC document *Conservation of mass in dissolving and precipitation*[3].

Outline requirements

- copper(II) sulfate 5-water (DANGER: Corrosive, harmful, irritant)
- 1 M ethanoic acid
- 0.01 M lead nitrate solution (not classified as hazardous but apply usual control measures for lead compounds)
- marble chip(s)
- 0.01 M potassium iodide solution
- sugar
- goggles
- carbonated drinks bottle
- micro-balance (cover second decimal place reading with insulation tape)
- small glass vials
- tap water

Outline method

1. Add the two substances separately to two glass vials, choosing one of the following:
   a. copper(II) sulfate-5-water and water
   b. sugar and water
   c. 0.01 M lead nitrate solution and 0.01 M potassium iodide solution.
2. Weigh and note the mass of both vials.
3. Combine both substances into one vial and stir to fully dissolve/mix.
4. Reweigh and note the mass of both vials.
5. For ethanoic acid and marble chips, add the reactants to a carbonated drinks bottle. Screw on the cap, and weigh and note the mass of the bottle. Allow the reaction to proceed for a minute, then release the pressure by carefully unscrewing the cap. Re-weigh and note the mass of the bottle.

Chapter 6: Rates of reaction and dynamic equilibrium

Observation 1: mass is conserved when we can see a change: something has changed – copper(II) sulfate 5-water dissolves into water forming a blue solution.

Observation 2: mass is conserved when a substance disappears – sugar dissolving into water forming a colourless solution.

Observation 3: mass is conserved when a new substance forms – lead iodide forms as a yellow precipitate.

Observation 4: mass is conserved when a gaseous substance forms and is trapped in the container – ethanoic acid and marble chips form carbon dioxide as a gas.

Microscale activity 6.4: Determining the equilibrium constant of a redox reaction

Ensure that full planning and risk assessment is carried out before attempting this activity.

Further details are available in the RSC *Practical Microscale chemistry: measuring an equilibrium constant on a microscale*[4].

Outline requirements

- 0.1 M iron(II) sulfate solution
- 0.02 M potassium thiocyanate solution
- 0.1 M silver nitrate solution (WARNING: Irritant)

- eye protection
- microburette
- small reaction vial
- syringes (1, 2, 5 cm^3)

Outline method

1. Measure 1 cm^3 each of silver nitrate and iron(II) sulfate solution into a vial.
2. Agitate the mixture to mix and leave to stand overnight.
3. Carefully transfer 1 cm^3 of the supernatant to a clean vial.
4. Fill the microburette with potassium thiocyanate solution.
5. Titrate the supernatant with the potassium thiocyanate solution. Thiocyanate ions precipitate the remaining silver ions, then form the dark red iron(III) thiocyanate complex – this is the end point.

$$Ag^+(aq) + Fe^{2+}(aq) \rightleftharpoons Ag(s) + Fe^{3+}(aq)$$

$$Ag^+(aq) + SCN^-(aq) \rightarrow AgSCN(s)$$

$$Fe^{3+}(aq) + SCN^-(aq) \rightarrow [FeSCN]^{2+}(aq) \text{ (dark red complex)}$$

Outline analysis

- Determine the initial concentrations of Ag^+ and Fe^{2+} in the mixture.
- Calculate the equilibrium concentration of Ag^+ from your titration data, hence the equilibrium concentration of Fe^{2+}.

- Hence calculate the equilibrium concentration of Fe^{3+}.
- Hence calculate the equilibrium constant, K_c.

This practical includes a solid product, which therefore does not appear in the equilibrium constant, making the analysis further challenging to students.

Teacher activities

Consider how the equilibrium constant of an esterification could be determined: for example, the synthesis of ethyl ethanoate from ethanol and ethanoic acid. Think about how a group of students could approach the equilibrium in their reactions from different starting points. Once the reaction has reached equilibrium, the solution can be analysed by titration of the remaining carboxylic acid with sodium hydroxide. Could you use microscale titration to analyse this solution? Consider using a strong-acid ion exchange resin as a heterogenous catalyst, which removes the complication of a homogenous catalyst being involved in the titration (see Chapter 10).

Consider how reversible reactions could be carried out on small scale. How could the dehydration/rehydration of copper sulfate be carried out at lower temperatures and in smaller quantities using a spirit burner and bottle top crucibles?

References

[1] http://science.cleapss.org.uk/Resource-Info/PP041-The-thiosulfate-acid-reaction-rate-and-concentration.aspx [Membership required] (Accessed August 2021)

[2] http://science.cleapss.org.uk/Resource-Info/PP141-The-thiosulfate-acid-reaction-rate-and-temperature.aspx [Membership required] (Accessed October 2021)

[3] https://edu.rsc.org/lesson-plans/conservation-of-mass-in-dissolving-and-precipitation-11-14-years/75.article (Accessed August 2021)

[4] https://edu.rsc.org/resources/measuring-an-equilibrium-constant-on-a-microscale/537.article (Accessed August 2021)

Chapter 7: Electrochemistry

'Hmm, I thought this microscale scale was for the less able!'
(A comment from a Head of Chemistry on seeing the electrolysis of copper chloride solution in a Petri dish, overheard by Bob).

Concepts being developed

Electrochemistry is a tricky topic, with many conceptual and technical challenges for the student, and even the teacher. Learning aspects by rote can provide intellectual crutches early on (e.g. acronyms such as PANIC, OILRIG[1]). Central to real understanding of the chemistry is to try and envisage what is occurring at the particle level, and consider the energy transfers involved. A Johnstone triangle (Figure 7.1) shows some further considerations.

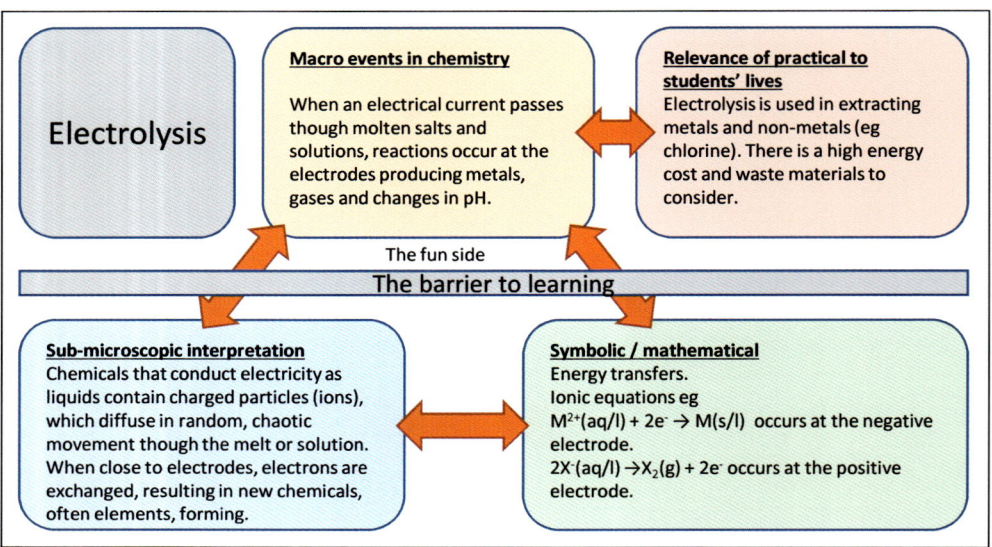

Figure 7.1: A Johnstone triangle on electrochemistry

When attempting to explain electrolysis, simplification can sometimes mask what is really happening. While it is common to say that the negative electrode will attract positive ions from the electrolyte, the actual influence of the electrode is only over a few nanometres. It is the movement of the solvated ions chaotically through the electrolyte that causes them to be close enough to be attracted.

Electrolysis tends to be tackled later on in chemistry curricula, and usually after electricity has been discussed in physics lessons. The flow of electrons in external circuits can be modelled in many different ways (e.g. ropes, heating circuits, trains) and can lead to misconceptions about the flow of charged particles within the electrolyte. Energy transfer is often spoken about differently between disciplines. Physics tends to use ideas of energy stores and transfers, while chemistry will often use ideas of enthalpy changes and energy cycles (Figure 7.2). Discussion of the models and ideas used by physics teachers in your department is always illuminating.

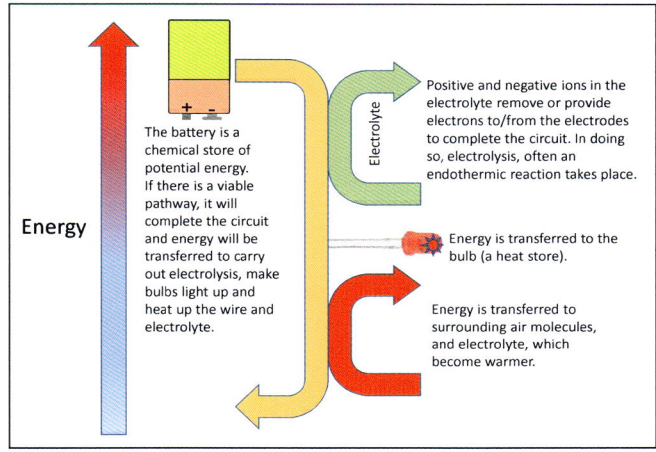

Figure 7.2: Diagrams used to bring together the idea of energy cycles and transfers

A possible teaching sequence for the electrolysis topics is:

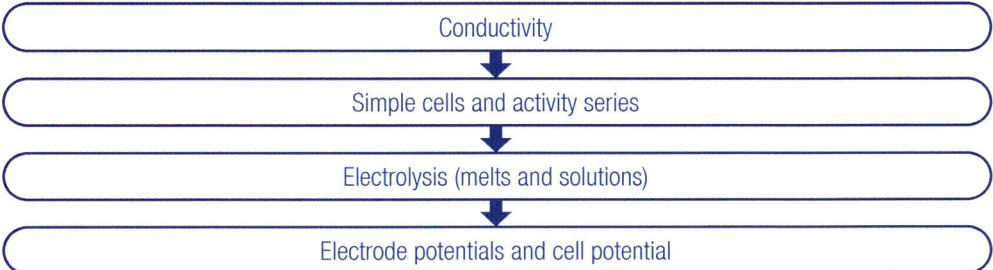

Conductivity of solutions allows us to discuss various aspects of chemistry with students, including the presence of ions in solution, production of industrially important chemicals, such as chlorine and aluminium, and electroplating.

The conductivity meter (see text box below) is a simple integrated device that allows conductivity investigations of a range of materials, including different solids (e.g. copper rods and glass) and substances in different forms (e.g. solid sodium chloride and sodium chloride solution).

Chapter 7: Electrochemistry

Microscale activity 7.1: Conductivity and ions in solution

Ensure that full planning and risk assessment is carried out before attempting this activity.

Outline requirements

- dropper bottle of water
- sodium chloride solid
- eye protection
- conductivity meter
- instruction sheet in plastic wallet
- wooden splint

Outline method

1. Make up an instruction sheet based on this outline method, and insert into the plastic wallet.
2. Place a few grains of salt near to where the puddle will be placed.
3. Make a puddle of water about 1.5 cm across on the instruction sheet.
4. Test the conductivity of the water by placing the probes in the puddle. The LED should not light up or at least be very dim.
5. Push a few crystals of salt to the edge of the puddle so that they just enter the liquid.
6. Hold the probes steady in the puddle.
7. Observe the LED over the next minute.

Expected outcome

In this procedure, there is no immediate light from the LED when the salt is added at the edge of the water. The salt dissolves and solvated ions are now free to diffuse through the liquid. As the ions encounter the electrodes, the circuit is complete as shown by the LED lighting up (Figure 7.3). Note that the solution will start to electrolyse producing hydrogen and chlorine gases.

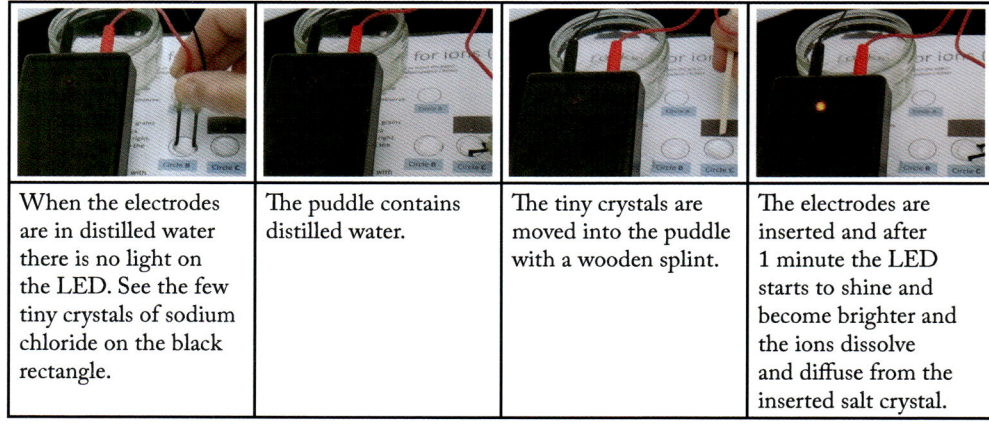

| When the electrodes are in distilled water there is no light on the LED. See the few tiny crystals of sodium chloride on the black rectangle. | The puddle contains distilled water. | The tiny crystals are moved into the puddle with a wooden splint. | The electrodes are inserted and after 1 minute the LED starts to shine and become brighter and the ions dissolve and diffuse from the inserted salt crystal. |

Figure 7.3: The diffusion of sodium and chloride ions through the puddle, which allows the current to flow indicated by the LED

Technical consideration – Conductivity indicator (ion detector)

The conductivity indicator consists of a 9 V battery, an LED, two 2 mm carbon fibre rods and a 300-500 Ω resistor (Figure 7.4). The device can be made up in a small box to protect the circuit. Full details are available in the CLEAPSS *Make-it guide, A conductivity indicator*[2].

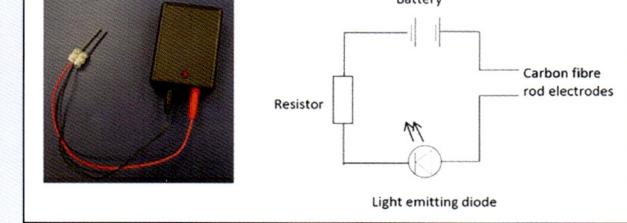

Figure 7.4: The conductivity indicator (the ion detector)

Students use batteries (two or more electrochemical cells linked together) in all manner of electrical devices. An understanding of how batteries work, and the environmental consequences of production and disposal is important. Research into batteries has led to a recent Nobel prize[3] and the move to battery-powered cars and even planes.

Many chemical reactions rely on the movement of electrons from one atomic environment to another. The electrochemical cell works by shifting energy from its chemical energy store, via the movement of electrons, to other energy stores. In doing so, useful work can be done, such as lighting bulbs or carrying out reactions like electrolysis.

Simple cells can be made using a minimum amount of metal materials. It makes for a useful investigation because, once the technique is shown, students can establish a 'league table' of activity using other metals and from demonstrations carried out by the teacher.

Microscale activity 7.2: Simple chemical cells

Ensure that full planning and risk assessment is carried out before attempting this activity.

Outline requirements

- Small pieces of
  - magnesium (1 cm strip of ribbon) (DANGER: Flammable)
- zinc
- iron (metal paper clips)
- copper
- 0.1 M sodium sulfate(VI) solution

- eye protection
- filter paper cut to fit Petri dish
- multimeter
- Petri dish

Understanding chemistry through microscale practical work 67

Outline method

1. Place the filter paper in the Petri dish.
2. Add drops of the sodium sulfate solution onto the filter paper until saturated (there shouldn't be puddles of the solution).
3. Place a piece of magnesium and copper on the filter paper.
4. Use the multimeter to determine the potential difference (DC voltage) between the two pieces of metal (Figure 7.5). Remove and replace the probes a few times to obtain consistent readings.
5. Repeat steps 3 and 4 with iron and zinc, and all other pair-wise combinations.
6. Demonstration – use small granules of calcium or lithium to show that magnesium isn't always the positive electrode.

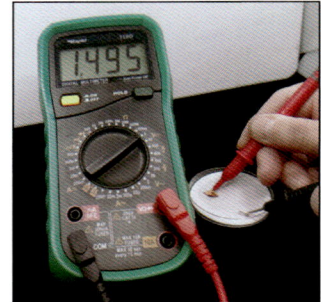

Figure 7.5: Measuring the cell potential of a magnesium/ion cell

Comparison of the measured potential difference allows an activity series relative to magnesium to be constructed. This can be compared with the series obtained from displacement reactions between metals and metal salt solutions. If lead is being used, the pieces should be handled with forceps. Aluminium foil can be tried, but tends not to give very stable readings. In general, factors such as presence of oxide layers, formation of insoluble salts and presence of nitrate ions can affect the readings.

Electrolysis

The process of using electricity to facilitate chemical reactions is known as electrolysis and the products are fundamental to our society. For example, aluminium containers and foils are made by the electrolysis of aluminium oxide dissolved in the molten mineral cryolite. Other metals, such as magnesium and lithium, are also obtained by electrolysis. Chlorine and sodium hydroxide, used in fighting harmful viruses and bacteria, are made by the chloralkali process involving the electrolysis of aqueous sodium chloride.

Electrolysis of aqueous chloride salts produces chlorine, a toxic gas. Carrying out larger-scale electrolysis experiments in poorly ventilated laboratories can cause severe breathing difficulties. These problems inspired the development of a microscale electrolysis set-up. Only a small volume of gas (about 6 cm^3) is produced and is both contained and largely used up in reactions in the Petri dish. The standard carbon electrodes were too wide and fragile to thread through holes in the side of the Petri dish. However, 2 mm-diameter carbon fibre rods used in making kite frames proved ideal. This robust modern material is difficult to break by students, but it can be cut with scissors.

Chapter 7: Electrochemistry

Microscale activity 7.3: Electrolysis of copper(II) chloride solution

Ensure that full planning and risk assessment is carried out before attempting this activity.

This is based on CLEAPSS Practical procedure PP059, *Microscale electrolysis of copper(II) chloride solution*[4].

Outline requirements

- dropper bottles of
  - 0.5 M copper(II) chloride
  - 0.5 M potassium bromide
  - 0.1 M potassium iodide
- blue litmus paper
- distilled water
- eye protection
- battery clip with crocodile clips
- carbon fibre electrodes (2)
- electrolysis cell (see text box below)
- low voltage DC supply set at about 9 V or a PP3 9 V battery

Outline method

1. Set up the electrolysis apparatus as shown in Figure 7.6.
2. Cover the set-up with the lid.
3. Connect the electrodes to the 9 V battery.
4. Observe the electrolysis in the central channel, the potassium halide solution drops, and the indicator paper.

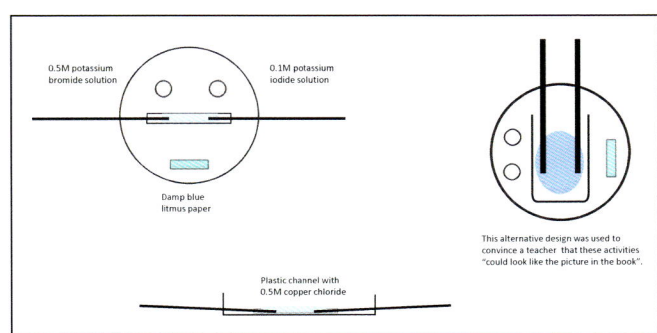

Figure 7.6: The set-up of the microscale electrolysis cell (left and middle). Right – an alternative set-up made to look like the electrolysis diagrams in textbooks

Expected outcomes

- Copper is seen forming at the negative electrode.
- Chlorine can be seen as bubbles at the positive electrode, and faintly smelled.
- The potassium halide solutions turn brown as bromine and iodine are displaced from solution.
- The litmus paper turns red and/or white as chlorine dissolves forming hydrochloric acid and chlorate(I) ions.

Understanding chemistry through microscale practical work

Chapter 7: Electrochemistry

> ### Technical consideration – Electrolysis cells
>
> Make holes on opposite sides of a Petri dish base (with a heated nail or cut out). Cut the shaft or bulb of a plastic pipette in half and to length and glue to the base in line with the holes. Cut 2 mm pieces of carbon fibre to length. A container can be made from corrugated plastic (e.g. Corriflute) or acrylic plastic using laser cutters (Figure 7.7). Liaise with your DT or Art department and see CLEAPSS *Make-it guide* GL163[5].

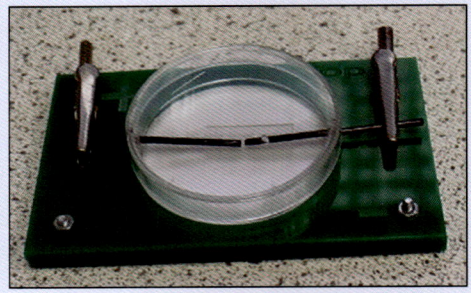

Figure 7.7: A microscale electrolysis apparatus

Electrolysis of a molten salt

Demonstrating the presence of ions in a molten salt using electricity is notoriously difficult. Salts with melting points below 600 °C are not common. Sodium chloride melts at over 800 °C. It can just be melted using a hot Bunsen flame, but borosilicate glass begins to soften at this temperature. Lead bromide is the traditional salt to use for electrolysis of molten salts, but the process needs to be carried out in a fume cupboard since the salt, lead and bromine are all toxic. All students tend to see is a bulb glowing or a meter needle moving. It is a triumph of the microscale approach that this important demonstration can be carried out as a demonstration on the open bench and projected onto a screen with a visualiser.

Microscale activity 7.4: Conductivity and electrolysis of molten salts

Ensure that full planning and risk assessment is carried out before attempting this activity.

This is based on CLEAPSS *Practical Procedure, Microelectrolysis of molten salts*[6].

Outline requirements

- solid lead bromide (DANGER: Harmful, serious health hazard)
- zinc chloride (DANGER: Harmful, corrosive)
- silver bromide
- lead chloride (DANGER: Harmful, serious health hazard)
- clean iron nail
- nichrome wire
- ethanol (DANGER: Flammable)

- eye protection
- 300-500 Ω resistor
- low voltage DC supply set at about 9 V or a PP3 9 V battery
- medium-wall borosilicate glass tubing (20 cm length, 6-7 mm diameter) bent, two thirds of the way along, at 90°
- red LED
- spirit burner
- wires and crocodile clips

Outline method

Lead bromide works best (despite the health/environmental concerns). Zinc chloride is very hydroscopic, silver bromide is photosensitive.

1. Add about 0.5 g of solid salt to the tube.
2. Clamp the glass tubing (on the longer arm) with the ends pointing upward.
3. Insert an iron nail (shorter arm) and a piece of nichrome wire into opposite ends of the tube as electrodes (carbon fibre electrode will decompose under the heat).
4. Complete the circuit by connecting a 9 V battery, 300-500 Ω resistor and an LED in series with the electrodes (Figure 7.8).
5. Heat the solid salt with a spirit burner until the solid is molten.
6. Observe the LED and the contents of the tube.

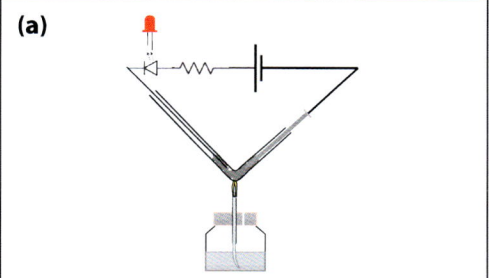 

Figure 7.8: The electrolysis set-up for molten salts (a) and production of bromine and lead from lead bromide (b)

The electrolysis of copper chloride solution and molten lead bromide yield ideal results, with the metal being deposited at the negative electrode whilst at the positive electrode the halogen is evolved.

Microscale activity 7.5: Determining standard cell potentials

Ensure that full planning and risk assessment is carried out before attempting this activity.

Determining standard cell potentials requires metal electrodes and solutions at 1.0 M, and saturated potassium nitrate salt bridges. This is usually carried out with beakers of solution and pieces of filter paper soaked in the potassium nitrate solution. A microscale alternative uses multiwell plates, and agar-based salt bridges.

Chapter 7: Electrochemistry

Outline requirements

- agar
- copper wire
- 1 M copper(II) sulfate solution (DANGER: corrosive and harmful)
- potassium nitrate(V) (DANGER: Oxidiser)
- zinc strip
- 1 M zinc sulfate solution (DANGER: Corrosive)
- goggles
- beaker
- silicone tubing, 3 mm
- hot water
- syringe, 2 cm^3
- scissors
- multiwell plate
- multimeter

Outline method

1. Melt 1 g of agar in 5 g potassium nitrate(V) in 50 cm^3 of hot water.
2. Draw the solution through a 3 mm diameter silicone tube using a syringe and allow the mixture to set.
3. Cut the salt bridge tube to 10 cm lengths.
4. Add 1 M copper(II) sulfate solution into one well, and 1 M zinc sulfate solution into an adjacent well in the multiwell plate.
5. Arrange the copper and zinc electrodes and the salt bridge into the wells.
6. Measure the electrode potential with a multimeter (Figure 7.9).

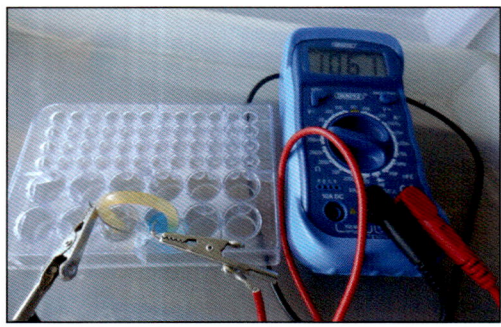

Figure 7.9: Measuring standard electrode potential of an electrochemical cell

Educational and technical Issues

- Two competing reactions often occur at the electrode, e.g. at the negative electrode, it is common to see the metal forming and hydrogen gas being evolved at the same time. In aqueous solutions, water itself is often involved in the electrolysis of many salt solutions and gases can be seen at both electrodes.

    $H_2O(l) + 2e^- \rightarrow H_2(g) + 2OH^-(aq)$ at the negative electrode (*reduction*).

    $H_2O(l) \rightarrow 2H^+(aq) + \tfrac{1}{2}O_2(g) + 2e^-$ at the positive electrode (*oxidation*).

- Altering the electrodes from carbon to another material alters the result. The surfaces of the various electrodes are different environments altering the reaction mechanisms pathways. This explains why, in the Hofmann Voltameter (Chapter 9), platinum electrodes are required to produce hydrogen and oxygen in a 2 to 1 volume ratio. With carbon electrodes, carbon dioxide, not oxygen, is often produced at the positive electrode. Iron cannot be used as the positive electrode in the electrolysis of molten lead bromide, because iron reacts with bromine at high temperatures.
- Textbooks and exam specifications often imply that as soon as the electrodes are added to an electrolyte, all the negative ions go to one and the positive go to the other. In fact, the charges on the electrodes are very short range (a few nm). It is diffusion and the random chaotic movement of the solvated ions that bring the particles close to the electrode surface so that a reaction is possible.
- There is common confusion over the terms 'cathode' and 'anode'. Reduction always occurs at the cathode (RedCat) and oxidation always at the anode (anOde). However, the charge of the electrodes change depending on whether the cell is electrochemical or electrolytic. Some curricula place more emphasis on electrolysis early on, so most students readily learn that anodes are positive and cathodes are negative. The reversal of charge when electrochemical cells are formed in 16+ courses then leads to much confusion. Focusing at the particle level helps a great deal, as well as considering what is happening with the electrons.

Teacher activities

- Consider how you could use the conductivity indicator to investigate the conductivity of covalently bonded substances – e.g., sucrose and glucose.
- Consider how the purity of different water samples could be investigated.
- Investigate how the nature of the electrode affects the reactions occurring. How does copper wire or iron (a paper clip) affect the reaction? Production of iron from iron(II) salt solutions can be tested for with a neodymium magnet.

Further reading

Professor Keith Taber (https://www.educ.cam.ac.uk/people/staff/taber/) has written extensively on how students (and teachers) can misunderstand some of the basic ideas. Bob always struggles with thermodynamics and quantum chemistry, and David has to go back to the books each year when sixth form electrochemistry comes around!

Teachers can find an extensive survey on Google scholar (https://scholar.google.co.uk/citations?user=mvYAZJUAAAAJ) and many of his articles are available on the Researchgate website.

Conceptual confusion in the chemistry curriculum: exemplifying the problematic nature of representing chemical concepts as target knowledge: https://link.springer.com/article/10.1007/s10698-019-09346-3 (an open access article)

Should we sacrifice Inquiry-Based Science Education (IBSE) in order to climb on PISA-rankings? Svein Sjøberg, Conference: 17th Nordic Research Symposium on Science Education

(NFSUN 2017) June 2017, Norway: https://www.researchgate.net/publication/315799796_Should_we_sacrifice_Inquiry-Based_Science_Education_IBSE_in_order_to_climb_on_PISA-rankings

References

[1] Common acronyms taught to students include 'Positive Anode, Negative Is Cathode' and 'Oxidation Is Loss (of electrons), Reduction Is Gain (of electrons)'.

[2] http://science.cleapss.org.uk/Resource-Info/GL166-Make-it-guide-a-conductivity-indicator.aspx [Membership required] (Accessed August 2021)

[3] https://www.nobelprize.org/prizes/chemistry/2019/press-release/ (Accessed August 2021)

[4] http://science.cleapss.org.uk/Resource-Info/PP059-Micro-electrolysis-of-copper-II-chloride-solution.aspx [Membership required] (Accessed August 2021)

[5] http://science.cleapss.org.uk/Resource-Info/GL163-Make-it-guide-microscale-electrolysis-apparatus.aspx [Membership required] (Accessed August 2021)

[6] http://science.cleapss.org.uk/Resource-Info/PP089-Microelectrolysis-of-molten-salts.aspx [Membership required] (Accessed August 2021)

Chapter 8: Further microscale activities

'Not long after I proposed the microscale cracking of hydrocarbons, a teacher, new to teaching but with many years of industrial experience, contacted me. It seems that she had got all of her sixth form students to do microscale cracking; they took 15 minutes with no mishaps. To her it was a revelation compared to the previous times that she had taught this lesson, when the practical took a whole lesson plus the added worry of explosions. She added: "You know what you are doing, don't you?" "Yes. Making it safe." "No, you have reduced the load on the short-term working memory. We encountered that in our teacher training. I could see it in the way the students did the practical with no fuss and I had the time to teach the chemistry in the same lesson." This began to make me think that there was more to this style of chemistry practical than just safety' (From Bob).

Some general concepts being developed

This chapter looks at a range of other activities and techniques that can help students develop their understanding of chemical concepts.

Identification of ions is a common technique and area of content to learn for many students. Simple identification tests for common ions based on precipitation can be easily carried out on plastic surfaces. Figure 8.1 shows an adaptation of a method from CLEAPSS Practical Procedure *Anion analysis* where the whole reaction is carried out in one puddle of water.

Under these conditions, carbonates, sulfates and nitrates do not produce precipitates helping with the identification of halides. Similar grids can be set up to test for:
- Carbonates. Use universal indicator solution and test any puddles that are alkaline by adding 1 M hydrochloric acid, which produces bubbles of carbon dioxide.

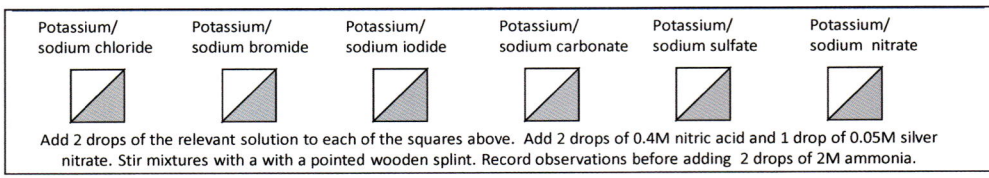

Figure 8.1: Identification of halide ions with silver nitrate solution (Ensure that full planning and risk assessment is carried out before attempting these reactions)

- Sulfates. Use 0.1 M barium chloride or nitrate solution. If a puddle shows a white precipitate then addition of 1 M hydrochloric acid will react with barium carbonate but not barium sulfate

A video with more information is available[1]. Significant time and effort is saved for students, teachers and technicians alike when carrying out activities like this on a microscale.

It is also possible to carry out tests on positive ions with similar grids. The required metal ions can be arranged horizontally using 0.1 M solutions and the test reagent, 0.4 M sodium hydroxide, arranged vertically. A more advanced procedure can be undertaken by testing with 2 M ammonia. A video with more information is available[2].

More advanced techniques can be carried out individually. For example, presence of the nitrate(V) ion can be demonstrated (activity 8.1) and the various reactions of transition metals (activity 8.2)

One of the most exciting aspects of microscale techniques is carrying out procedures, often not thought possible in the school setting, that enhance students' experience and understanding of chemistry. Visible evidence for colourless gases can be achieved by liquifying ammonia, and solvated electrons can be shown by reaction with lithium (activity 8.6). A different reaction for demonstrating reversible reactions is possible with the synthesis of the interhalogen compounds (activity 8.7).

If schools base their organic preparative chemistry on the 50 cm^3 flask with the B14/23 ground glass joints, often called semi-micro, then that is small enough and there is no cost saving in going smaller, although kits do exist. But catalytic cracking can be performed in such a way that the dangers of suck-back are significantly reduced (activity 8.8). Free-radical polymerisation, which requires a fume cupboard for long refluxing, can be successfully achieved (activity 8.9). There is also a synthesis of propene (activity 8.10) to be used as an alternative to using cyclohexene as the model compound for investigating the chemistry of alkenes (activity 8.11). This has the added bonus of being used to show hydrogenation.

Chapter 8: Further microscale activities

Microscale activity 8.1: Identification of nitrate(V) ions

Ensure that full planning and risk assessment is carried out before attempting this activity.

Identification of positive ions was introduced in Chapter 2, and can be taken further with quick tests for negative ions[3]. Identification of less often tested ions, such as nitrate(V), can be carried out successfully on a microscale.

Outline requirements

- aluminium foil or granules
- bromothymol blue indicator solution (WARNING: Flammable)
- 0.5 M potassium nitrate(V) solution
- red litmus paper
- 1 M sodium hydroxide solution (DANGER: Corrosive)
- water
- goggles
- forceps
- Petri dish, small
- pipette
- small reaction vessel (e.g. empty blister pack section)

Outline method

1. Place a damp piece of litmus paper, a few drops of bromothymol blue solution and the reaction vessel separately in the base of the Petri dish.
2. Add 5-6 drops of potassium nitrate(V), 5-6 drops of sodium hydroxide solution and a small amount of aluminium granules in the reaction vessel.
3. Place the lid on the Petri dish and observe (Figure 8.2).

Figure 8.2: Nitrate(V) is reduced to ammonia by aluminium in alkaline solution. The ammonia turns the red litmus paper blue and the bromothymol blue from yellow/green to blue

Microscale activity 8.2: Chemistry of iron and other transition metals

Ensure that full planning and risk assessment is carried out before attempting this activity.

Transition metals have general properties, such as variable oxidation states, coloured compounds, complex ions, paramagnetism and catalytic activity. However, each metal has their own peculiarities. This might be a blessing for the research/industrial chemist but is a struggle for the student coping with exams. The technique of adding reagents to a plastic surface with instruction underneath is ideally suited to the study of transition metals.

This activity looks at the chemistry of iron(II). Students can then repeat this activity using hydrated iron(III) nitrate to compare and contrast the observations between iron(II) and

Chapter 8: Further microscale activities

iron(III) irons (Figure 8.3). This activity can be adapted to study the chemistry of other transition metals such as copper(II).

Outline requirements

- solid hydrated iron(III) nitrate (WARNING: Irritant)
- dropper bottles of:
  - distilled water
  - 0.4 M sodium hydroxide (WARNING: Irritant)
  - 0.1 M potassium hexacyanoferrate(II)
  - 0.1 M potassium hexacyanoferrate(III)
  - 0.1 M potassium thiocyanate
- "10 - 12 vol", ~1 M , ~6% hydrogen peroxide
- 1 M sodium thiosulfate solution
- eye protection
- pipettes
- pointed wooded splint stirrer
- reaction worksheet in plastic wallet
- scissors

Outline method

1. Place a small spatula of solid iron(II) sulfate and iron(III) nitrate on the plastic wallet.
2. Add about 20 drops of distilled water to each solid to make solutions – stir to dissolve.
3. Add 2-3 drops of the following solution in the five reaction circles of each solution.
   (a) 0.4 M sodium hydroxide,
   (b) 0.1 M potassium hexacyanoferrate(II),
   (c) potassium hexacyanoferrate (III)
   (d) 0.1 M potassium thiocyanate,
   (e) "10 - 12 vol" hydrogen peroxide
4. Make a puddle of water about 1 to 1.5 cm in diameter. Using two wooden splints move a crystal or two of iron(III) nitrate and potassium thiocyanate into the puddle from either end.
5. To 2 or 3 drops of iron(III) nitrate solution, add drops sodium thiosulfate solution. Stir to watch the changes of colour. Once the addition of more drops of sodium thiosulfate produces no colour add 0.4 M sodium hydroxide.

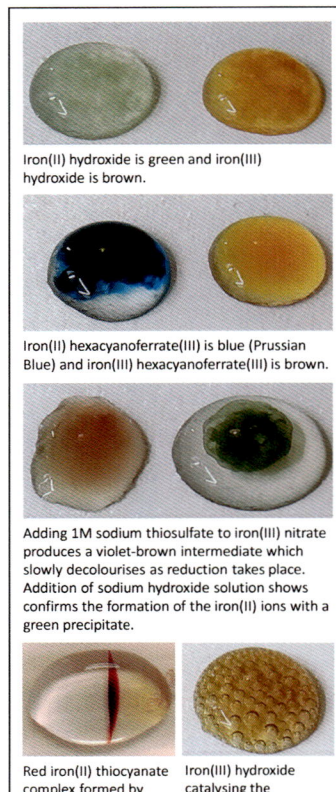

Iron(II) hydroxide is green and iron(III) hydroxide is brown.

Iron(II) hexacyanoferrate(III) is blue (Prussian Blue) and iron(III) hexacyanoferrate(III) is brown.

Adding 1M sodium thiosulfate to iron(III) nitrate produces a violet-brown intermediate which slowly decolourises as reduction takes place. Addition of sodium hydroxide solution shows confirms the formation of the iron(II) ions with a green precipitate.

Red iron(II) thiocyanate complex formed by adding the solids from each side.

Iron(III) hydroxide catalysing the decomposition of 10 vol hydrogen peroxide to form bubbles of oxygen.

Figure 8.3: Reactions of iron(II) and iron(III) ions

Chapter 8: Further microscale activities

Microscale activity 8.3: The chemistry of rusting

Ensure that full planning and risk assessment is carried out before attempting this activity.

Rusting costs the economy millions of pounds per year, and students will be familiar with rust on bicycles and cars. This activity goes beyond the rudimentary experiment carried out early in school chemistry and looks at how indicators can be used to show up some of the chemistry occurring, and how different metals can act as sacrificial metals.

Outline requirements

- copper foil
- ferroxyl indicator (0.02% potassium hexacyanoferrate(III), 0.2% sodium chloride, 0.002% phenolphthalein in distilled/deionised water)
- magnesium ribbon (DANGER: Flammable)
- zinc foil
- eye protection
- iron nails (4) or wire
- paper
- Petri dish, 9 cm
- sandpaper

Outline method

1. Prepare four pieces of sanded iron wire about 2 cm long.
2. Separately wrap a piece of copper foil, zinc foil and magnesium ribbon around pieces of the iron wire.
3. Place the piece of metal in the base of a Petri dish or on a plastic wallet.
4. Add ferroxyl indicator to each piece of metal, ensuring that part of the iron is exposed to the air in each case.
5. Observe the corrosion of the metals over time (Figure 8.4).

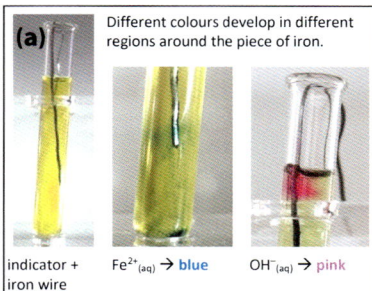

Figure 8.4: The chemistry of rusting (a) ferroxyl indicator showing the different reactions occurring (Source: CLEAPSS) (b) sacrificial protection of iron

Expected results

The ferroxyl indicator shows the presence of both the iron(II) ions forming (Figure 8.3) as a blue precipitate, and hydroxide ions forming (a pink colouration). The copper has no protective effect on the iron, whereas both zinc and magnesium corrode in preference to the iron. The chemistry can be taken further by considering the electrode potentials of the reactions involved in rusting, including the oxidation of iron to iron(II) and iron(III) and the formation of the hydroxides.

Declan Fleming wrote an excellent article on rusting for *Education in Chemistry*, which provides further details[4].

Microscale activity 8.4: Catalysis of reaction between zinc and hydrochloric acid

Ensure that full planning and risk assessment is carried out before attempting this activity.

This activity looks at a familiar displacement reaction where multiple observations can be made and discussed. The subtleties of electrochemical cells can be observed when simple combinations of metal wire are exposed to acid. Intriguing observations can be made, which are useful to help students think deeply about how particles are moving in the system.

Outline requirements

- 1 M hydrochloric acid
- copper wire
- zinc wire
- eye protection
- pipettes
- plastic wallet with paper inside

Outline method

1. Add a few drops of hydrochloric acid onto the plastic wallet.
2. Place a piece of zinc in the acid drop – observe.
3. Wrap a piece of copper wire around another piece of zinc wire – place this in a separate drop of acid – observe (Figure 8.5).

Figure 8.5: An electrochemical cell formed by copper and zinc in acid

Expected results

As expected, zinc reacts with acid to produce hydrogen gas. However, the reaction is slow. Chemists, when making hydrogen, add copper(II) sulfate solution to coat the zinc with a copper metal to act as a catalyst. Remarkably, it also works with copper wire and zinc wire twisted together. By focusing on the observations, students and teachers can see that the bubbles of hydrogen come from the copper wire, yet the solution does not go blue – it is the zinc that dissolves.

Chapter 8: Further microscale activities

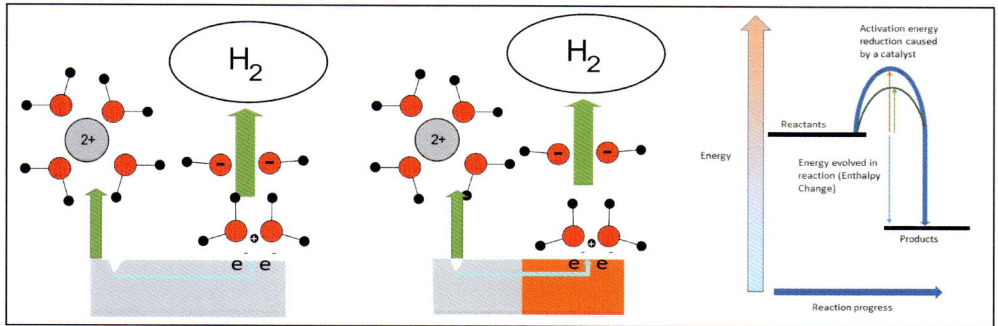

Figure 8.6: The mechanism and energetics of the reaction of zinc with acid

This simple experimental set-up should allow students plenty of 'mental space' to consider the reaction mechanism. The electrons in zinc metal must be moving into the copper wire. Is this a more favourable pathway? Certainly, copper is a better conductor of electricity than zinc. When the electrons reach the surface of the copper wire, they react with the solvated protons (from the acid) that are colliding with the copper surface producing hydrogen. The electrons are falling into a positive 'hole', namely the (solvated) hydrogen ions. They do this because the potential energy of the system is reduced via a convenient kinetic pathway. Simultaneously, zinc atoms on the surface of the metal are removed by solvation (Figure 8.6).

Microscale activity 8.5: Detailed observations of a displacement reaction

Ensure that full planning and risk assessment is carried out before attempting this activity.

This reaction, mentioned in Chapter 2, shows bubbles forming on magnesium ribbon in freshly made iron(II) sulfate, raising the excellent question from students, *'Is this supposed to happen?'*. These side reactions are distractions from the topic that the teacher is trying to teach, and the effects are not often mentioned in revision or textbooks. It is important to stress that the reaction is 'not going wrong'.

The magnesium obtains a coating of iron (black) and is attracted to the magnet (red object). That is 'normal'. However, iron also catalyses the reaction between magnesium and water, releasing hydrogen and hydroxides ions; the precipitate of iron(II) hydroxide (green), air oxidising to iron(III) hydroxide (brown) is beginning to appear, which is often not expected. The movement of electrons from magnesium to iron is responsible for this and is a competing reaction.

Outline requirements

- solid hydrated iron(II) sulfate (WARNING: Irritant)
- magnesium ribbon, 5 mm length and cut in half lengthways as well, if possible (DANGER: Flammable)
- eye protection
- bar magnet
- plastic wallet
- pointed wooded splint stirrer

Outline method

1. Add a few crystals of iron(II) sulfate onto the surface of the plastic wallet.
2. Add water to make a small puddle and stir.
3. Place a piece of magnesium into the drop, pushing it down with the stirrer.
4. Observe the reactions and different products formed (Figure 8.7).
5. Test for the presence of iron with the magnet.

Figure 8.7: The many products of the reaction between magnesium and acidified iron(II) sulfate (the red object on the right is a bar magnet)

Microscale activity 8.6: Liquifying ammonia and the reaction with lithium

Ensure that full planning and risk assessment is carried out before attempting this activity.

One of the triumphs of the microscale approach is doing an activity not often thought possible in a school laboratory. Acid solutions contain the hydrogen ion, or better described as the hydrated or solvated proton. The addition of lithium to liquid ammonia forms the beautiful blue solution of the solvated electron.

Outline requirements

- concentrated ammonia solution (DANGER: corrosive)
- lithium (DANGER: Flammable, causes burns)
- propanone (DANGER: Highly flammable, irritant)
- solid carbon dioxide (WARNING: Gas is asphyxiant, can cause frostbite)

- goggles
- access to a fume cupboard
- beakers (2)
- bung
- digital thermometer, reading to -76 °C
- filter paper
- medium wall glass tubing (20 cm long, 0.6–0.7 mm diameter) bent into a U-tube that fits into the beaker
- side arm boiling tube
- silicone tubing

Outline method

1. Work in a fume cupboard, wear goggles and gloves.
2. Set up the side arm boiling tube connected via silicone tubing to the glass U-tube.
3. Place the U-tube in a propanone/solid carbon dioxide mixture in a beaker.
4. Add concentrated ammonia solution to the side arm boiling tube and bung the tube.
5. Observe liquid ammonia forming in the U-tube.
6. Add a small piece of lithium (remove the oil on filter paper) into the U-tube.
7. Observe the formation of blue solvated electrons (Figure 8.8).

Expected results

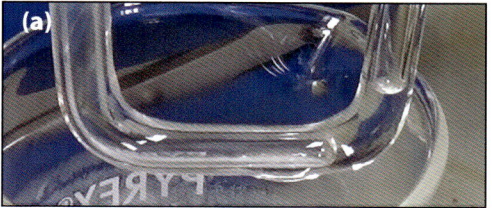 

Figure 8.8: (a) Formation of liquid ammonia (b) Solvated electrons when lithium dissolves in ammonia

Microscale activity 8.7: Synthesis of and equilibrium between interhalogens

Ensure that full planning and risk assessment is carried out before attempting this activity.

Interhalogen compounds are rather exotic for schools, but the yellow iodine trichloride is relatively easy to make and exhibits a very obvious reversible reaction with iodine monochloride.

$$ICl(l) + Cl_2(g) \rightleftharpoons ICl_3(s)$$

This activity uses anhydrous copper(II) chloride that produces dry chlorine when heated.

Outline requirements

- anhydrous calcium chloride (WARNING: Irritant)
- copper(II) chloride-2-water (WARNING: Harmful, irritant)
- iodine (WARNING: Harmful)
- mineral wool

- eye protection
- access to a freezer
- glass Petri dish
- hot plate
- hot water from kettle
- medium wall glass tubing (20 cm long, 0.6-0.7 mm diameter) bent into a U-tube that fits into the beaker
- silicone tubing
- small test tube
- three-way taps (2) to end as stoppers to U-tube (blu-tac can work)

Chapter 8: Further microscale activities

Outline method

1. Prepare anhydrous copper(II) chloride by warming copper(II) chloride-2-water on a glass Petri dish on a hot plate.
2. Connect together the following:
   a. small test tube containing anhydrous copper(II) chloride
   b. silicone tube containing anhydrous calcium chloride between pieces of mineral wool
   c. U-tube containing iodine (Figure 8.9).
3. Heat the copper(II) chloride gently with a Bunsen flame.
4. Observe the formation of iodine chloride (brown liquid) and iodine trichloride (yellow solid).
5. Seal the end of the U-tube with stoppers. Place the U-tube in hot water/freezer to demonstrate the reversible reaction between the iodine chlorides.

Expected results

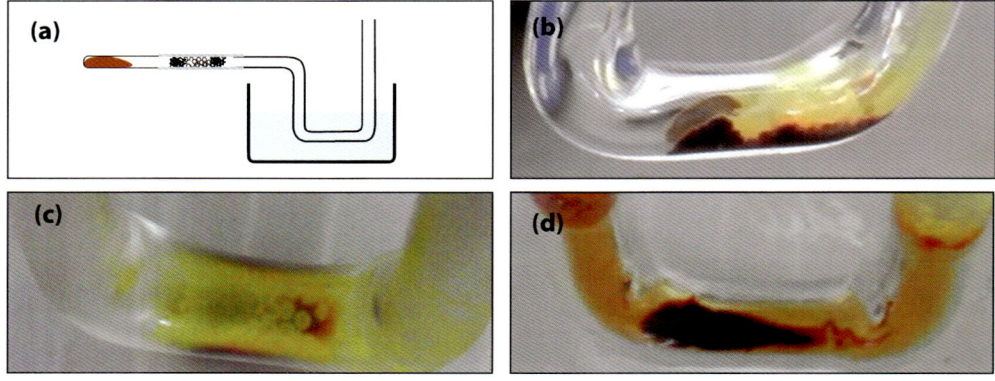

Figure 8.9: Formation of interhalogen compounds (a) Diagram of the experimental set-up (b) iodine chloride and iodine trichloride form in the U-tube (c) and (d) the reversible reaction can be demonstrated by placing in a freezer/hot water

Microscale activity 8.8: Catalytic cracking of hydrocarbons

Ensure that full planning and risk assessment is carried out before attempting this activity.

This is based on the CLEAPSS *Practical Procedure PP060, Catalytic cracking of hydrocarbons*[4].

This method has several advantages over the traditional large-scale method, which often calls for a broken ceramic pot as the catalyst. The reduced scale significantly reduces the hazards from suck-back, removes the need for Bunsen burners, and reduces the release of a strongly-smelling product. This method also combines the production and testing of the product in one step.

Chapter 8: Further microscale activities

Outline requirements

- 0.002 M bromine water
- ethanol (DANGER: Highly flammable)
- mineral wool
- paraffin oil (DANGER: Flammable, serious health hazard, irritant)
- 0.002 M potassium manganate solution
- powdered aluminium oxide
- eye protection
- 90° delivery tube
- fine metal stiff wire to push the wool down
- Reaction vessel (12 × 75 mm reaction tube or sealed glass Pasteur pipette)
- silicone tubing
- spirit burner

Outline method

1. Set up the reaction vessel (Figure 8.10):
   a. mineral wool soaked in paraffin oil
   b. powdered aluminium oxide
   c. silicone tube connected to 90° delivery tube.
2. Place the end of the delivery tube in 0.002 M bromine water or 0.002 M potassium manganate(VII) solution.
3. Heat the aluminium oxide with a spirit burner, then move the flame to the junction between the mineral wool and aluminium oxide.
4. Dispose of the used reaction tubes in glass waste.

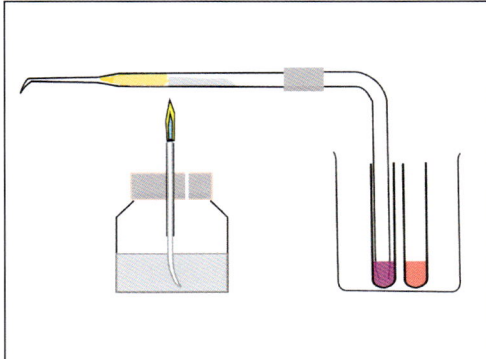 

Figure 8.10: Microscale cracking. A broken and then sealed glass Pasteur pipette is used in the diagram. However, a small test tube can also be used

Expected results

Bubbling will be seen through the test solutions. If catalytic cracking is occurring, the test solutions will decolourise. The colour change with potassium manganate(VII) solution is clearer, although bromine water is usually quoted as the test reagent. The main reason for this reaction not working is not using enough catalyst and/or the catalyst not being hot enough. As a heterogenous surface catalyst, a large amount is required (similar to catalytic converters in motor vehicles). Industry and research catalysts can include zeolites, which have an open cage structure (hence a large surface area).

Microscale activity 8.9: Additional polymerisation

Ensure that full planning and risk assessment is carried out before attempting this activity.

Outline requirements

- dilauroyl peroxide (DANGER: May cause fire if heated)
- methyl methacrylate (DANGER: Highly flammable, irritant, lachrymatory vapour)
- mineral wool
- goggles
- access to boiling water
- glass pipette (plastic ones will soften in methyl methacrylate)
- gloves
- hammer
- newspaper
- sealed Pasteur pipette (reaction pipette)

Outline method

1. Add a small mass of lauryl peroxide to a sealed Pasteur pipette.
2. In a fume cupboard, add methyl methacrylate to the reaction pipette using a glass pipette.
3. Plug the end of the reaction pipette with mineral wool.
4. In the open laboratory, place the reaction pipette in a beaker of just-boiled water.
5. Maintain the temperature at 50-60°C (or leave until the next lesson).
6. Observe the formation of the polymer (opaque at higher temperature, turns clear at room temperature) (Figure 8.11).
7. The polymer can be removed from the reaction pipette by wrapping in newspaper and CAREFULLY breaking the glass with a hammer (wear gloves and eye protection).

Expected results

The reaction works well with styrene. Longer polymerisation times at lower temperatures leads to improved polymer formation.

Chapter 8: Further microscale activities

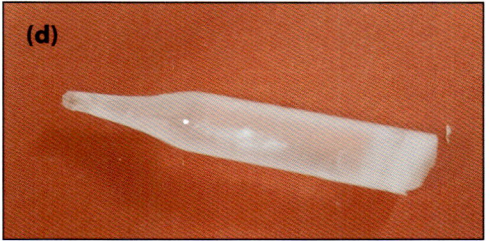

Figure 8.11: Synthesis of addition polymers (a) Sealed glass Pasteur pipettes are filled with about 0.05 to 0.08 g of dilauroyl peroxide and filled up to half-way with methyl methacrylate (b) Boiling water is added to the beaker and kept at about 50 to 70°C for about 90 minutes (it can be left to the next day) (c) As polymerisation proceeds, the liquid becomes viscous (d) On cutting open the glass, the solid polymer is revealed in the shape of the pipette

Microscale activity 8.10: Synthesis of propene

Ensure that full planning and risk assessment is carried out before attempting this activity.

This activity is a useful alternative to the usually lengthy and low-yield preparation of cyclohexene (itself a foul-smelling substance). The use of three-way taps and stoppers with the syringes makes gas handling simple and safe.

Outline requirements

- 0.002 M bromine water
- mineral wool
- 0.002 M potassium manganate solution
- powdered aluminium oxide
- propan-2-ol (DANGER: Flammable, irritant)

- eye protection
- blunt syringe needle
- reaction vessel (12 × 75 mm reaction tube or sealed glass Pasteur pipette)
- S-bend delivery tube
- syringes, 60 cm³ (2)
- three-way tap

Understanding chemistry through microscale practical work   87

Chapter 8: Further microscale activities

Outline method

1. Set up the apparatus as shown in Figure 8.12.
2. Fill the vertical syringe with water by connecting the two syringes via the three-way tap and pulling the plunger of the horizonal plunger.
3. Set up the reaction vessel in the same way as in activity 8.8, replacing paraffin oil with propan-2-ol.
4. Heat the catalyst with the spirit burner (no need to heat the propan-2-ol directly, given the low boiling point).
5. Observe the propene collected as a gas in the vertical syringe.
6. Extract the propene from the vertical syringe into another syringe via the three-way tap.
7. Test the unsaturated nature of the substance by bubbling through bromine water/potassium manganate(VII) solution and the smoky flame produced when burnt (see activity 8.11).

Expected results

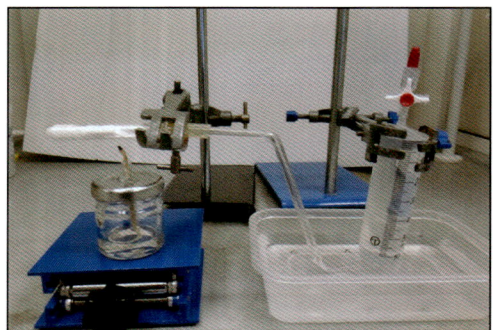 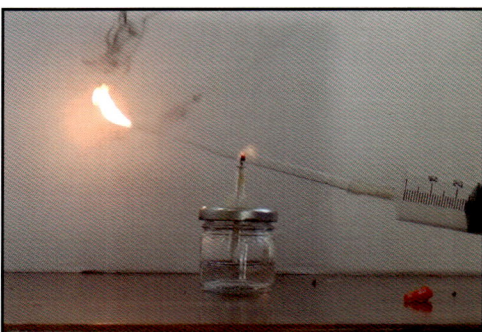

Figure 8.12: Set-up for producing propene (left) and the smoky flame produced when combusted (right)

Technical considerations – three-way taps

The three-way taps with the stoppers are extremely useful in dealing with gases (Figure 8.13). They are used in medical applications to transfer liquids and gases through various lines. Larger versions are used in research studying reactions in an inert atmosphere or vacuum.

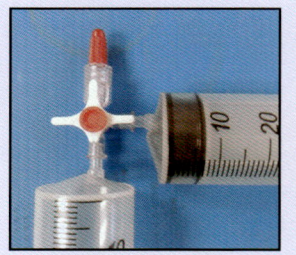

Figure 8.13: A three-way tap

Microscale activity 8.11: Hydrogenation of propene

Ensure that full planning and risk assessment is carried out before attempting this activity.

Outline requirements

- source of propene (see activity 8.10) (DANGER: Extremely flammable)
- source of hydrogen (DANGER: Extremely flammable)
- 10% palladium catalyst on carbon
- eye protection
- catalyst tube (see Box on page 90)
- silicone tubing
- syringes, 60 cm³ (2)

Outline method

1. Set up two syringes separately containing 25 cm³ of propene and 25 cm³ of hydrogen.
2. Connect the syringes via silicone tubing to a catalyst tube (Figure 8.14).
3. Carefully push the gases back and forward between the syringes to mix the gases and allow the hydrogenation reaction.
4. Compare the smokiness of the flames of propane and propene (Figure 8.15).

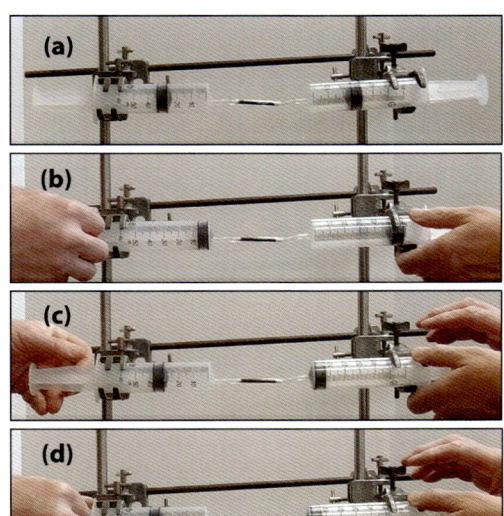

Figure 8.14: Reaction of propene and hydrogen over a catalyst

(a) 25 cm³ of propene is in the left hand syringe and 25 cm³ of hydrogen in the right hand syringe. The 10% palladium on carbon catalyst is between them

(b) Propene is passed over the catalyst by pushing the plunger of the left-hand syringe and pulling the plunger of the right-hand syringe (hydrogen). There is 50 cm³ of gas in the right-hand syringe

(c) Pushing the mixture back over the catalyst causes a reduction of volume in accordance with the equation $C_3H_6(g) + H_2(g) \rightarrow C_3H_8(g)$

(d) After 2 more passes, the volume of gas in the right hand syringe reaches 25 cm³, providing evidence for Gay-Lussacs's Law of Combining Volumes of Gases. Touch the catalyst tube carefully, it is hot

Expected results

- The volume of gas will reduce from 50 cm³ overall to 25 cm³ overall as two moles of gases combine to one mole of gas, following Gay-Lussac's Law of Combining Volumes.
- The catalyst tube will get hot due to the exothermic nature of the hydrogenation reaction.
- This reaction is a model for industrially important reactions such as hydrogenation of vegetable oils to make fats and making cleaner burner hydrocarbons.

Chapter 8: Further microscale activities

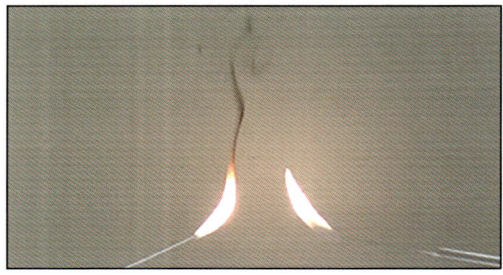

Figure 8.15: Comparison of the flames of propene (left) and propane (right)

Technical consideration – catalyst tubes

This method requires a catalyst of palladium on carbon or aluminium oxide (Figure 8.15). With carbon, mineral wool plugs are used to prevent the carbon from 'blowing away' during the push me-pull me technique with syringes. 1 g of the catalyst from suppliers might be expensive, but it is enough for several tubes.

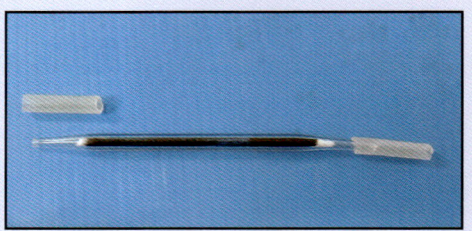

Figure 8.16: A catalyst tube

Teacher activities

- Design reaction worksheets for the ion tests required by your exam board specification.
- The microscale preparation of copper sulfate crystals can produce crystals within 20 minutes from copper(II) oxide. See https://youtu.be/L1mI4IHQJsc. Make an integrated worksheet for the activity. Notice the filtration and heating techniques, which have not been covered in the book in detail.

Further reading

By accessing the website https://www.chemedx.org, teachers can access the *Journal of Chemical Education* on https://www.chemedx.org/search/result/Journal. Selected articles are available as open access.

References

[1] Subscribers to CLEAPSS can access the Practical Procedure PP109 – *Anion analysis: Microscale*. Teachers in Scotland can access this via SSERC. See also https://youtu.be/gMqyP1J2KS0 which is an open resource.

[2] Subscribers to CLEAPSS can access the Practical Procedure PP098 – *Testing for positive ions: reactions with hydroxides and ammonia*. Teachers in Scotland can access this via SSERC. https://www.youtube.com/watch?v=ziB461SHyLo&t=313s

[4] Fleming, D. (2016) Nail Corrosion Demonstrations *Education in Chemistry* - https://edu.rsc.org/exhibition-chemistry/nailing-corrosion-demonstrations/2000054.article (Accessed November 2021).

[4] Subscribers to CLEAPSS can access the Practical Procedure PP100, *Test reagents for negative ions*. Teachers in Scotland can access this via SSERC.

[5] http://science.cleapss.org.uk/Resource-Info/PP060-Catalytic-Cracking-of-Hydrocarbons.aspx [Membership required] (Accessed October 2021)

Chapter 9: STEM connections

'A favourite is the microscale Hofmann voltameter, which provides a simple yet attention-grabbing starter for a lesson on hydrogen fuel cells. The hydrogen and oxygen produced can be collected in one syringe and bubbled through soap solution held in a Petri dish. When ignited with a lit splint, the bubbles burn with a surprisingly loud crack, illustrating the large amounts of energy that can be released from this "perfect" reaction'
(Catherine Smith, Head of Science, Hinckley Academy and John Cleveland Sixth Form Centre).

Concepts being developed

A central aspect of providing a practical education is the experience that it gives students of applying their creativity and hands-on skills to solving problems. Developing these competences can support students on the path to further study and careers in the fields of Science, Technology, Engineering and Mathematics (STEM). The UK economy is worth about £3 trillion per annum, with industry and agriculture accounting for about 20%. The 2017 UK Government Industrial Strategy[1] indicated that 40% of employers reported shortages of STEM graduates as being a barrier to their recruitment[2]. Following the Gatsby Career Benchmarks[2], particularly Benchmark 4 – linking curriculum learning to careers, all teachers have a responsibility to support students in their career development, and chemistry teachers are well placed to do this.

The development of many of the microscale apparatus and techniques has come from analysing a particular practical or technique, identifying issues that need to be solved, and being creative in the employment of materials to use. For example, how do we solve the problem of not having enough water baths to allow students to complete temperature-controlled reactions? Solution: make a water bath from a plastic container (Chapter 6). For example, how do we minimise the cost of breakages when combusting magnesium in porcelain crucibles? Solution: make crucibles out of readily available bottle caps (Chapter 4).

Chapter 9: STEM connections

This trial and improvement approach to how science and engineering is done in the real world is perhaps at odds with the impressions that many students (and teachers) can have of these fields. Certainly, having a firm grasp of the underlying concepts is vital to being a successful scientist or engineer. However, much of the development of new materials or instruments comes from time in the lab, trying things out, with 'failed' experiments leading to insights and refinement of ideas. Giving students a flavour of this during their science education will provide a more authentic education and the tacit knowledge that cannot be gained from reading a textbook or watching a demonstration. This will help prepare them for higher education study and careers in STEM fields, including biomedical sciences (Figure 9.1).

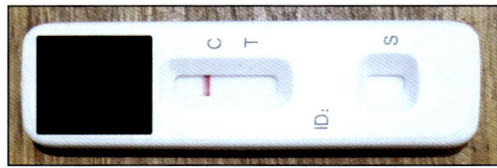

Figure 9.1: A lateral flow test for SARS-CoV-2 antigens – microscale chemistry by the million during the Covid-19 pandemic

Microscale activity 9.1: Investigating concentration using a DIY colorimeter

Ensure that full planning and risk assessment is carried out before attempting this activity.

Outline requirements

- brass object
- copper foil
- concentrated nitric(V) acid (DANGER: Oxidiser, toxic, corrosive)

- goggles
- Access to a fume cupboard
- 500 Ω resistor
- 9 V battery
- crocodile clips
- detector LED (Use an IR LED)
- digital multimeter
- emitter LED (colour depends on the colour of the solution)
- LEGO® blocks
- plastic cuvette
- wires

Outline method

1. Find the accurate mass of some copper foil weighing between 0.7 and 0.8 g. In the fume cupboard, add it to a minimum volume of concentrated nitric acid (about 5 ml) and, after it has dissolved, make the solution up to 100 cm^3 with distilled/deionised water in a volumetric flask. Label this solution A.
2. Repeat Step 1 with a brass item.
3. Make up solutions in vials containing 2 cm^3 of A with 8 cm^3 of water, 4 cm^3 of A with 6 cm^3 etc.
4. Assemble the LEGO® bricks into a cuvette holder, with technic bricks either side for the LEDs (Figure 9.2a).

Chapter 9: STEM connections

5. Insert the emitter LED into one technic brick. Assemble a series circuit joining the 9 V battery, the switch, the resistor and the LED.

6. Insert the detector LED into the opposite technic brick. Connect the LED to the multimeter and set to measure DC voltage.

7. Find which colour LED emitter produces the largest range of voltages in the detecting LED.

8. Fill the cuvette with deionised water and measure the 'blank' reading on the multimeter.

9. Find the voltage reading of each vial (Figure 9.2b).

10. Construct a concentration vs absorbance graph (Figure 9.2c) and assess how closely the data fits to the Beer-Lambert law $A \propto log(V_o/V)$ where A is the Absorbance, Vo is the voltage reading for pure water and V is the voltage reading for a coloured solution.

11. Now find the voltage of the solution made from brass and find the mass of copper in the brass and hence its percentage.

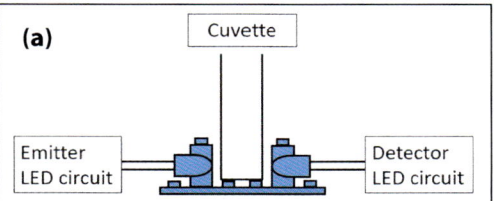

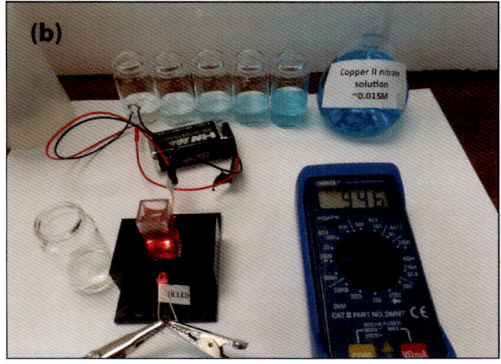

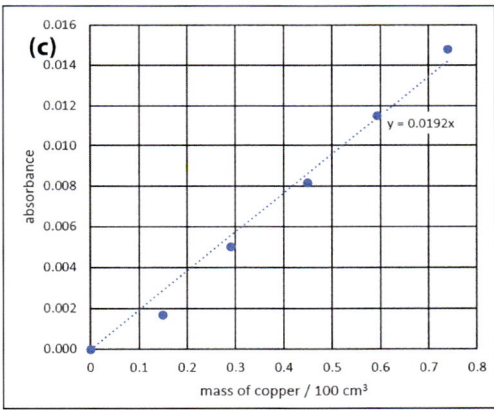

Figure 9.2: (a) Schematic of the DIY colorimeter
(b) Practical set-up using the DIY colorimeter
(c) Close correlation between experimental data and Beer-Lambert Law

Once the DIY colorimeter has been constructed, students can investigate reactions that show a colour change over the course of the reaction. For example:
- iodination of propanone;
- iodine clock reactions, such as the Harcourt-Essen and Landolt reactions;
- displacement reaction between copper sulfate and zinc;
- reaction of thiosulfate and acid;
- acid-base neutralisation using indicators; and
- bicarbonate indicator colour change in photosynthetic systems.

Understanding chemistry through microscale practical work

It can also be used to find or compare the concentration of coloured solutions such as:

- comparing the intensity of colours in various carbonate drinks.

Investigating the properties of gases

The Hofmann voltameter was invented in 1866 and allowed the productions of oxygen and hydrogen gas by the electrolysis of water. Platinum electrodes are used, and a small amount of ionic substance, such as sulfuric acid or sodium sulfate, is used to increase the conductivity of the water. The traditional glass apparatus found in schools resembles the original design. They are rather expensive and prone to breaking due to the fragile nature of the glass joints, sticking of the stopcocks and corrosion around soldered joints connecting the electrodes. Also, only expensive platinum electrodes will deliver the 2:1 ratio of hydrogen to oxygen that textbooks claim.

A microscale version was suggested in the RSC *Microscale Chemistry* publication, using plastic pipettes and hot glue. However, this was tricky to assemble and prone to leakages.

An improved version was designed by CLEAPSS and a *Make-it guide* is available[3]. Two syringes are arranged over a small Petri dish with the platinum electrodes inserted through the dish. Three-way gas taps are used so that the gases being produced can be collected and analysed.

Microscale activity 9.2: Investigating the electrolysis of water

Ensure that full planning and risk assessment is carried out before attempting this activity.

Outline requirements

- bromothymol blue solution
  (WARNING: Flammable)
- Petri dish filled with a soap solution
- 0.1 M sodium sulfate(VI) solution

- eye protection
- low voltage DC supply set at 9 V or 9 V PP3 battery
- microscale Hofmann voltameter
- syringes: 5 cm^3, 10 cm^3 and 20 cm^3

Outline method

1. Carry out the electrolysis of sodium sulfate solution using the microscale Hofmann voltameter (Figure 9.3).
2. Observe the relative volumes of oxygen and hydrogen being produced – consider whether this fits with the expected ratio.
3. Repeat the electrolysis using bromothymol blue in the electrolyte – consider why the solution next to the negative electrode is turning blue (alkali) and the positive electrode is turning yellow (acid).
4. Repeat the electrolysis but without the dye and syringes. Use a pH probe (and meter) to examine the pH over the electrodes.
5. Repeat the electrolysis:

a. Extract the hydrogen produced at the negative electrode (about 5 cm^3), transfer the gas to an inverted test tube (volume about 20 cm^3) and insert a lighted splint. The familiar 'pop' will confirm hydrogen.
b. Extract the oxygen produced at the positive electrode and carry out the confirmation gas test with a glowing splint.
6. Repeat the electrolysis and collect a 1:2 ratio of oxygen:hydrogen in a syringe. Bubble this mixture through soapy water and light the bubbles with a burning splint.
7. Insert into circuit a multimeter, reading current (A). Measuring the time (in seconds) for the formation of 5 cm^3 of hydrogen allows you to calculate the quantity of electricity passed (Q) and can be linked to Faraday's First Law of electrolysis.

Technical consideration

One advantage of the microscale version is the use of low hazard 1 M sodium sulfate as the electrolyte in place of 1-2 M sulfuric acid in the traditional equipment. Now the formation of acidic and alkaline areas around the two electrodes becomes apparent, agreeing with the half equations shown in text books. There is some evidence that a small amount of ozone is produced around the positive electrode as well as oxygen. This can decolourise the indicators if used for long periods.

Figure 9.3: A microscale Hofmann voltameter

Chromatography

Chromatography is a vital analytical technique used in a wide range of academia and industries. Students will usually experience this technique early in their secondary (or even primary) education when separating pen inks or food colourings and then, again, further on when separating out reaction mixtures during organic synthesis.

The principles underlying these simple chromatographies are the same for advanced chromatography found in universities and industrial laboratories. These include high performance liquid chromatography (HPLC), gas-liquid chromatography (GLC), ion-exchange and size-exclusion chromatography. Forensic science makes wide use of these techniques, including analysis of explosive residues, presence of drugs in victims, and presence of trace DNA and fibre evidence at crime scenes. Chromatography is also widely used by the medical profession in detecting drug metabolites, identifying markers for disease and quality control during pharmaceutical development.

Chromatography is often carried out on cellulose filter papers using water soluble dyes from ink pens. Careful selection of pens can give reasonable separation. However, the spots tend to be quite smeared, and it is difficult to measure retention factor values convincingly.

Small-scale separation of hydrophobic mixtures on TLC plates in organic solvents tends to give much better separation and can be carried out in only a few minutes. Suitable mixtures include the chlorophyl from leaves, inks from biro pens or lipstick.

You need to buy 20 × 5 cm silica plates on a plastic backing. Cut these into 5 × 1.1 cm plates. These are expensive, but the results are rewarding. You should purchase plates that fluoresce at 254 nm allowing sample identification with UV light (care must be taken due to the presence of UV-C light).

Microscale activity 9.3: Chromatography on thin-layer chromatography plates

Ensure that full planning and risk assessment is carried out before attempting this activity.

This is based on CLEAPSS *Practical procedure PP056, Thin-layer chromatography of plant pigments*[4].

Outline requirements

- anhydrous sodium sulfate(VI)
- extraction solvent (2:3 v/v mixture of ethyl ethanoate:propanone) (DANGER: Highly flammable, harmful)
- running solvent (5:2:2 v/v cyclohexane:ethyl ethanoate: propanone) (DANGER: Highly flammable, harmful)
- soft dark green leaves (e.g. spinach)

- eye protection
- blunt forceps
- fine paint brush
- small glass vial
- small tube (e.g., Eppendorf, or test tube)
- scissors
- TLC plate to fit vial

Outline method

1. Cut up about 2 cm^2 of leaf into very small pieces and place in the tube.
2. Add a small amount of sodium sulfate and 5 drops of extraction solvent.
3. Mix and crush the mixture with the forceps for about 2 minutes.
4. Mark the baseline (in pencil) on the TLC plate about 1 cm from the bottom.
5. Paint a small drop of the leaf mixture supernatant onto the baseline and allow to dry. Repeat 3-4 times ensuring a small final circle of mixture.
6. Add about 0.5 cm of running solvent into the vial and carefully place the TLC plate in the vial.
7. Cover the vial with the cap.
8. Observe the separation until the solvent front is about 0.5 cm from the top of the plate.
9. Remove the plate, mark the solvent front and allow the solvent to evaporate (Figure 9.4).
10. Measure the distance between the baseline and the solvent front, and between the baseline and the spots. Calculate R_f values.

Chapter 9: STEM connections

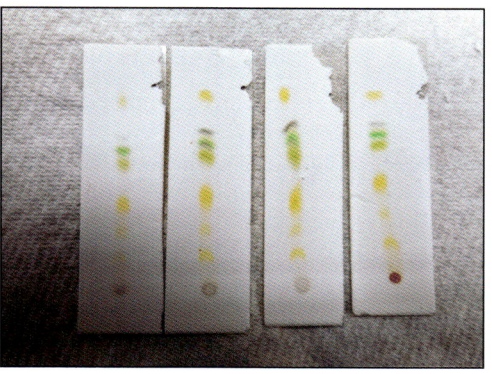 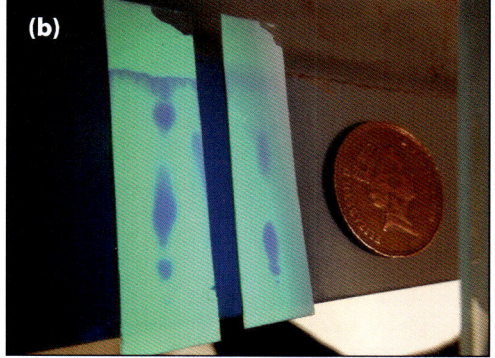

Figure 9.4: Microscale chromatography on TLC plates (a) The TLC of chloroplast extracts from pondweed, a dark geranium leaf and a light geranium leaf and the tradescantia leaf. Note the dark anthocyanin spot, which does not move (b) The TLC of colourless extracts from drugs viewed under UV-C light

Technical considerations

A great deal of the development of the microscale kits relies on the time and skills of technical staff in schools. CLEAPSS provides a wide range of *Make-it guides*, but the actual initial construction takes time and collaboration between science technicians and D&T technicians.

Many of these devices have come about through experimentation in the laboratory by people just giving something a try. Feedback from colleagues is essential, alongside trial with students and teachers to see whether the devices work for their educational purposes.

Social media and bulletin boards can help with this development. A quick picture on Twitter or a short video on *YouTube* usually yields lots of feedback and (mostly) constructive criticism. Demonstrating new ideas at conferences is also useful, since this can allow people to try out prototypes hands-on.

The advent of 3D printing and cheap microelectronics, such as the Arduino board, has increased the availability of custom-made apparatus and programmable sensors in secondary education. Our physics and D&T colleagues tend to be much greater users of these technologies and collaboration with them often yields interesting advances. These ideas are suitable for extended studies or student research.

Chapter 9: STEM connections

Neodymium magnets

Another new material available for school chemistry is the powerful neodymium magnet. The attraction of salts such as iron(III) nitrate and manganese(II) sulfate in Eppendorf tubes to these magnets (Figure 9.5a) provides evidence of the unpaired electrons in electronic structure of ions[5]. The magnets are even strong enough to attract a soap bubble of oxygen on the surface of water. Up to now, it has only been possible to demonstrate this with liquid oxygen[6]. The paramagnetic effects are also apparent in complexes (Figure 9.5b) and precipitates[6]. Even more remarkable is the effect on ions during electrolysis (Figure 9.5c). The magnets are placed under the electrodes and the products of electrolysis begin to circle around the electrodes.

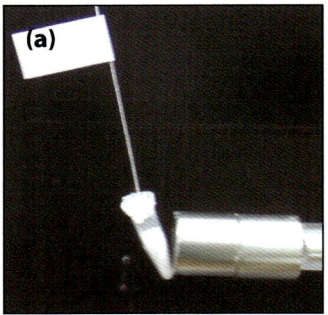 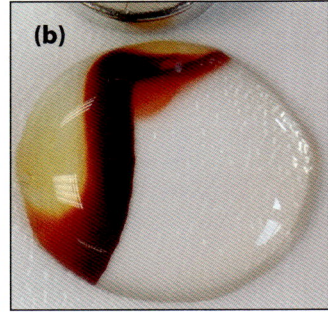 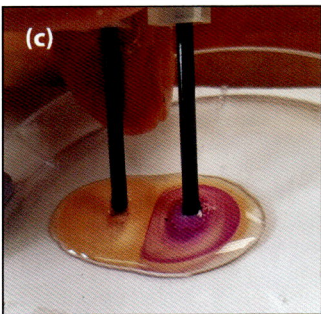

Figure 9.5: Examples of the effect of a magnetic field on (a) iron(III) nitrate crystals (b) the iron(III) thiocyanate complex and (c) the electrolysis of sodium sulfate solution containing universal indicator.

Teacher activities

- Consider how a Tic-Tac® box can be used as an alternative to plastic cuvettes in a DIY colorimeter.
- Consider how to include TLC earlier into secondary education schemes of learning and how this could help develop the careers education of the students. Forensic science is often a popular career choice for those interested in STEM degrees, aided by its central role in police drama shows, such as *NCIS*.
- Consider how chromatography is currently being used in our exploration of the solar system, for example, the search for life on Mars by the NASA Perseverance Rover.
- Many pieces of equipment can be made use 3D printing and laser cutting devices in conjunction with D&T departments and could become student projects. Examples include:
  - the electrolysis of copper chloride equipment;
  - the conductivity indicator;
  - the colorimeter;
  - syringe and test tube holder;
  - a microscale heater and stirrer; and
  - a lagged microscale calorimeter.

Further reading

CLEAPSS GL174 *Make-it guide – a simple colorimeter*: http://science.cleapss.org.uk/Resource-Info/GL174-Make-it-guide-a-simple-colorimeter.aspx [Membership required]

A guide on carrying out simple and inexpensive chromatography: http://science.cleapss.org.uk/Resource-Info/PS067n-Chromatography.aspx [Membership required]

Aspirin – A curriculum resource for post-16 chemistry and science courses, (2nd edition), Royal Society of Chemistry (2003), a useful context in which to use thin-layer chromatography. https://edu.rsc.org/resources/aspirin-book/56.article (Accessed October 2021)

Microfluidics takes the miniaturisation to even smaller levels. For more information, see https://aip.scitation.org/doi/10.1063/1.5096030, which surveys education-related articles. Also see https://www.researchgate.net/publication/263949278_Microfluidics_for_High_School_Chemistry_Students

References

[1] https://www.gov.uk/government/publications/industrial-strategy-building-a-britain-fit-for-the-future (Accessed June 2021)

[2] https://www.economicmodelling.co.uk/wp-content/uploads/2018/12/STEM-Report_vWEB.pdf (Accessed August 2021)

[3] http://science.cleapss.org.uk/Resource-Info/GL196-Make-it-guide-microscale-Hofmann-voltameter.aspx [Membership required] (Accessed October 2021)

[4] http://science.cleapss.org.uk/Resource-Info/PP056-Thin-layer-chromatography-of-plant-pigments.aspx [Membership required] (Accessed October 2021)

[5] https://www.youtube.com/watch?v=9-TD2Hbd71g (accessed November 2021)

[6] https://www.youtube.com/watch?v=jJmduBDr8Y8 (accessed November 2021)

Chapter 10: Sustainability

'Microscale experiments provide an invaluable tool to safely get that 'Wow! Eureka' moment. The experiments such as burning magnesium allow deeper insight into the underlying scientific knowledge. In the PGCE science practical days we introduce a range of microscale experiments and highlight how they act as a teaching tool to promote deeper scientific understanding and discussion amongst our students' (Rob Campbell, Senior Lecturer in Science Education, St Mary's University, Twickenham, London).

Education in Sustainable Development

I, Bob, remember that in my early days in Nottingham, the atmosphere was full of smoke from burning coal, and smells from the coke ovens and the glue factory. The top decks of the buses were full of people smoking cigarettes and coughing all the time. Pea-green coloured fogs and smogs were very common.

Students in previous generations will have learned, by rote, about important industrial processes of the time – such as steelmaking (blast furnace), sulfuric acid manufacture (the Contact process) and the use of coal both as a fuel and feedstock of industrial chemicals. However, there was little consideration of the impact on the environment or on the health of the people working in such industries.

For the industrial processes studied now, such as fractional distillation, the Haber process and electrolysis for extracting metals, there is a greater awareness of environmental impact as well as health and safety considerations. In addition, students may study some of the regulations imposed by governments to try to control some of the impact. How much meaning students take from this learning is open to debate.

Significant effort has been expended over recent decades in increasing the understanding of the impact of science on our society through the education system. During the 1980s, the Association for Science Education (ASE) (a UK-based subject association) published booklets in a general studies course called *Science in Society*. *Twenty-first Century Science*[1] and *Salters'*

Chemistry[2] present chemical ideas and practical skills relating to modern-day applications of chemistry and current research. The United Nations General Assembly proclaimed a 'Decade of Education for Sustainable Development' (ESD) in 2005, but teachers all over the world struggled to include ESD in an already overloaded curriculum. Jegstad and Sinnes (2015)[3] suggested five areas of teaching through an ESD agenda (Table 10.1).

Content	Prescribed by syllabus/specification.
Context	Prescribed by syllabus/specification. Also, gather information from present-day news, newspapers, television, books and social media, which the teacher should attempt to be familiar with ('Do you know that there are gold nanoparticles in every flow test taken for Covid-19?'). The importance of keeping science teachers up-to-date with modern progress in science is recognised by our learned societies, who try to relate to modern issues though teacher journals[4].
Chemistry's distinctiveness and methodological character	Relies on observations through practical work, to enhance understanding via the three corners of the Johnstone Triangle and through real-life issues. *'By consciously reflecting on green chemistry principles and the nature of chemistry, pupils may increase their understanding of the scientific process in chemistry in particular and in science in general.'*[3]
ESD competences	Emphasising the complex inter-relationship within the traditional three strands of science (biology, chemistry, physics), where aspects of chemistry reflect on effects on the environment, for example, mining for rare-earth elements for use in modern electronics and the waste from such devices. It then also reflects on other 'systems', such as economics, health provision: in other words, 'systems thinking'.
Lived ESD	'Involves both the role of the teacher in creating a friendly and safe learning environment (characterised by ESD principles) and the importance of a sustainable school culture. Lived ESD is emphasised as a part of the general education in schools but is also transferable to the lives the students live outside school.'[3]

Table 10.1: Five areas of teaching about sustainable development

CLEAPSS came to a similar conclusion that *'Green Chemistry Principles should be implicit in teaching and practical work in the first place, rather than relying on examination specifications'*[5].

There is a movement in the USA promoting Green Chemistry in Education[6]. Sustainability in chemistry is now accepted as a vital aspect of chemistry[7].

Interestingly, there are now Government initiatives being addressed by the Royal Society of Chemistry (RSC)[8]. In a vision for 2030, the RSC states that *'The UK education system, through school, higher and further education, and professional development, is training knowledgeable professionals and informed and scientifically literate citizens who understand the nature, value and importance of chemicals in products. Citizens appreciate how chemicals in our environment are managed sustainably, by using science and technology and circular economy models'*.

Chapter 10: Sustainability

Is Microscale chemistry green?

There are twelve guiding principles to green chemistry[9], seven of which are relevant to microscale techniques (Table 10.2).

1	Preventing waste	By using small amounts and volumes of chemicals, there is an obvious reduction in waste.
2	Atom economy	Addition, condensation and rearrangement reactions tend to have high atom economy reactions, compared with elimination and substitution reactions.
3	Using less hazardous chemical synthesis	This is also a requirement under UK Health & Safety COSHH Regulations.
4	Using safer solvents	Moving away from extremely flammable solvents, such as ethoxyethane, and environmentally toxic solvents, such as the organohalogens, is an established policy in UK science education.
5	Energy efficiency	Moving to less reliance on carbon-based fossil fuels is possible by using well-designed ethanol spirit burners or electrical hot plates. The Bunsen burner is used only when temperatures above 500 °C are required.
6	Catalysis	Using benign/reusable/heterogenous/selective catalysts is a fundamental requirement. An example is replacing sulfuric acid as an acid catalyst and using ion-exchange resins (see activity 10.3)
7	Inherently safer chemistry for accident prevention	Microscale chemistry reduces the reliance on fume cupboards, reduces the possibility of glass explosions after suck-back, and reduces the occurrence of large fires.

Table 10.2: Relevant principles of green chemistry in microscale chemistry

Cleaning pipettes

In microscale techniques, great use is made of the plastic transfer pipettes. They are incredibly cheap at 2p to 3p each. The dilemma is, if they have been used and contaminated, should they be thrown away or washed? This can be the subject of an investigation by the students and an example of 'Lived ESD'.

Investigation

0.02 M potassium manganate(VII) solution
1 M hydrochloric acid
0.1 M zinc sulfate solution (WARNING: Irritant)
0.4 M sodium hydroxide solution (WARNING: Irritant)

Indicators (WARNING: (Highly) Flammable)
Eye protection
Transfer pipettes
Conductivity indicator
pH probe and meter

An experiment ends with pipettes with droplets of chemical solutions adhering to the plastic sides. 'A wash' is defined as *"sucking" pure water into the pipette, inverting it so the water goes into the bulb and expelling the water into a beaker for waste'*.

Some tasks and questions

1. If the student is supplied with pipettes that have been used with 0.02 M potassium manganate(VII) solution, how many washes does it take to remove any traces of the purple liquid that have been left in the tube?
2. How can we check that a pipette is free from contaminant if colourless 1 M hydrochloric acid had been used and then washed in the above manner? (The method is in this book!)
3. How can we check that a pipette is free from contaminant if colourless 0.1 M zinc sulfate had been used? (The method is in this book!)
4. A student group uses 5 transfer pipettes for an activity. There are 12 groups of students in the class. Should the pipettes be thrown away into the waste bin? Who would be cleaning these pipettes? How long would that job take? Is it a good use of teacher or technician time to wash out these pipettes? How much money are you throwing away? Should pipettes be recycled? There are five forms of 12 groups of students doing this experiment in the school; how does this affect the calculation? There are about 6000 secondary schools doing this activity in the UK; how does this affect the calculation?
5. How can you dry your pipettes? Do they need to be dried? How would you use a slightly wet pipette in a new lesson with, say, a fresh solution of sodium hydroxide?

The simple solution to all this is that, if students cleaned their five pipettes at the end of the activity, then all of these questions are redundant. This can be applied to life in general; each of us is responsible for the waste we generate: 'Lived ESD'.

Some answers (but not all)

1. You can drop a puddle of the washings onto a plastic sheet and look for the colour.
2. You can drop a puddle of acid/alkali washings onto a plastic sheet and add 1 drop of an indicator (activity 2.1).
3. You can drop a puddle of acid/alkali or salt solution washings onto a plastic sheet and test with a conductivity indicator (activity 7.1).
4. If a pipette is damp with water and it is required for use with another reagent, then suck up the reagent solution and expel into a waste beaker a couple of times.
5. You can use ethanol to do a final washing. Ethanol evaporates at a faster rate than water (Why? The molecules are 2.5 times as heavy). It does not matter too much if the ethanol becomes diluted with traces of water. However, would it matter if a 0.4 M sodium hydroxide solution becomes 0.38 M because of a tiny amount of water in the pipette? It would matter, of course, if it were to be used for a quantitative procedure such as a titration.

Depolymerising PP and PE

What happens to polythene when used? It takes up to 1000 years for the average plastic bag to naturally decompose. When you consider the billions of plastic bags used every year, this issue becomes large. Answers can be found here[10]. With unwanted higher heavy fractions of oil, the molecules are broken down into smaller more useful molecules by a process called cracking (Chapter 8). Could the same process be carried out on waste polythene (Figure 10.1)? While alkenes are produced, is the yield, energy and catalyst use worth this depolymerisation or is it better to granulate the waste and recycle it into new products?

Figure 10.1: Packed polythene granules at the end of a sealed Pasteur pipette and aluminium oxide, heated with a spirit burner. Alkene gases being collected in an inverted syringe

Acid rain

Acid rain[11] was one of the big environmental talking points of the 1980s. It affected large swathes of woodland in Scandinavia, Canada, USA and Scotland. The problem was caused by sulfur dioxide and nitrogen oxides in the atmosphere dissolving in water. Both gases can occur naturally, sulfur dioxide from volcanoes and nitrogen oxides from electrical discharges. However, gases from the burning of fossil fuels such as coal, and the emission from vehicle exhausts, were added by humans.

Oxygen in the atmosphere can oxidise the gases to sulfur trioxide and nitrogen dioxide, both of which, on dissolving in water, form strong acids. The resultant decrease in the pH of rainwater affected both trees and fish stocks in lakes. This chemistry can be investigated with activities 10.1 and 10.2.

Chapter 10: Sustainability

Microscale activity 10.1: Investigating sulfur dioxide chemistry

Ensure that full planning and risk assessment is carried out before attempting this activity.

Outline requirements

- sodium metabisulfite (DANGER: Harmful, corrosive)
- universal indicator paper
- dropping bottles of
  - 0.1 M barium chloride solution
  - 1 M hydrochloric acid
  - 10 vol hydrogen peroxide
- tap water
- eye protection
- micro spatulas
- Petri dishes, 5.5 cm (2)
- small reaction vessel (e.g. a 'well' from a tablet blister pack such as throat lozenges)

Outline method

1. Set up a reaction vessel with a tiny mass of sodium metabisulfite (a few crystals) inside a Petri dish (Figure 10.2).
2. Set up two test drops in the Petri dish – one of tap water, one of tap water and hydrogen peroxide.
3. Add 1 drop of hydrochloric acid to the sodium metabisulfite and place the lid on the Petri dish.
4. After 1-2 minutes, test the pH of the two drops of water. Compare with a sample of tap water.
5. Repeat the method using 0.1 M barium chloride solution instead of tap water.

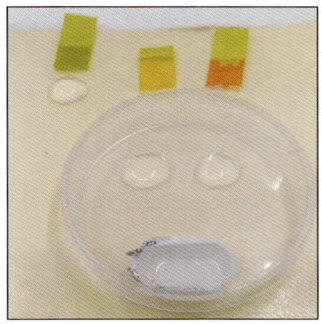 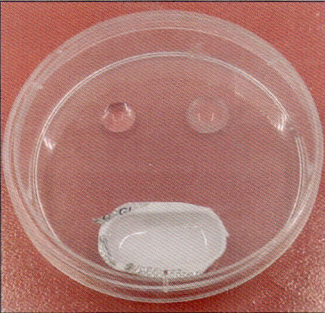

Figure 10.2: Experimental set-up for investigating sulfur dioxide chemistry (a) Testing pH (b) testing for sulfate ions

Expected results

Sulfur dioxide is a toxic gas, but microscale techniques allow the chemistry to be observed outside of the fume cupboard. In tap water, sulfur dioxide forms sulfuric(IV) acid (sulfurous acid), lowering the pH of the solution (left hand drop). The hydrogen peroxide simulates the oxidising agent present in the atmosphere, leading to sulfur trioxide and sulfuric(VI) acid formation, and a lower pH (right hand drop).

When barium chloride solution is used, the right hand drop turns cloudy, indicating the presence of sulfate ions, so confirming the presence of sulfuric(VI) acid.

Chapter 10: Sustainability

Microscale activity 10.2: Investigating the effects of nitrogen oxides on sulfur dioxide

Ensure that full planning and risk assessment is carried out before attempting this activity.

Outline requirements

- sodium metabisulfite (DANGER: Harmful, corrosive)
- sodium nitrite (nitrate(III)) (DANGER: Oxidiser, toxic)
- dropping bottles of
  - 0.1 M barium chloride solution
  - 1 M hydrochloric acid
- eye protection
- micro spatulas
- Petri dishes, plastic 5.5 cm (2)
- 2 x small reaction vessel (e.g. a 'well' from a tablet blister pack such as throat lozenges)

Outline method

1. Set up two reaction vessels separately with tiny masses of sodium metabisulfite and sodium nitrate(III) (a few crystals) inside a Petri dish (Figure 10.3).
2. Set up one test drop in the Petri dish of barium chloride solution.
3. Add 1 drop of hydrochloric acid to the reaction vessels and place the lid on the Petri dish.
4. Observe the barium chloride drop.

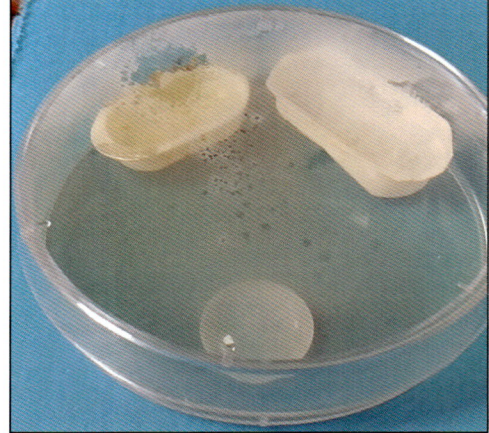

Figure 10.3: Experimental set-up for investigating sulfur dioxide chemistry in the presence of nitrogen oxide

Expected results

Nitrogen monoxide is formed by lightning and internal combustion engines. Nitrogen monoxide is then oxidised by oxygen in the air to nitrogen dioxide (the brown gas in Figure 10.3), and forms nitric acid when dissolved in water. Nitrogen dioxide oxidises sulfur dioxide to sulfur trioxide, allowing the formation of sulfuric(VI) acids in water, tested for with the barium chloride.

Catalysts

Catalysts are an important area of green chemistry. Finding recyclable heterogenous, highly selective catalysts that work below 100 °C in water is the focus of much research. This is why there is a lot of interest in phase transfer catalysts and ion-exchange resins. These are specially formulated thermoplastics (such as polystyrene) with attached organic functional groups on the surface. They have moved in and out of exam specifications for over 50 years. They are essential in removing hardness from water and are used in dishwashers where they are regenerated by adding salt. However, they can also be used in making esters, including aspirin, in the place of concentrated sulfuric or phosphoric acid (homogenous catalysts).

Microscale activity 10.3: Making esters with heterogeneous catalysts

Ensure that full planning and risk assessment is carried out before attempting this activity.

Outline requirements

- anhydrous sodium carbonate (WARNING: Irritant)
- butan-1-ol (DANGER: Flammable, irritant, corrosive)
- glacial ethanoic acid (DANGER: Flammable, corrosive)
- Strong acid resin (e.g. Amberlite® HPR1100)
- water
- goggles
- beaker of cold water
- filter paper and funnel
- hot plate
- small glass vial

Outline method

1. Add about 0.1 g of the resin to the vial (Figure 10.4).
2. Add about 10 drops each of ethanoic acid and butan-1-ol to the vial.
3. Place the vial on a hot plate until the mixture boils.
4. Use tongs to hold the vial in a beaker of cold water to cool the mixture.
5. Add about 10 drops of water and about 0.5 g of sodium carbonate to the mixture to neutralise any remaining acid.
6. Cautiously smell the mixture.
7. Recover the resin by filtration.

Figure 10.4: A sample of a strongly acidic resin used to catalyse esterification reactions

Teacher activities

- Compare the use of strong ion exchange resins against phosphoric or sulfuric acid as catalysts in the preparation of aspirin.
- Go to CLEAPSS Guide GL191, *Alternative water and suction pumps* and set up a recirculating water pump for distillation. What is the temperature rise in the water during one such trial?

Further reading

Go to https://www.beyondbenign.org/ which is an American website on education in green chemistry.

Explore the contribution made by the RSC on https://www.rsc.org/new-perspectives/sustainability/

Explore the UK Green Chemistry Centre of Excellence on https://www.york.ac.uk/chemistry/research/green/

References

[1] https://www.ocr.org.uk/qualifications/gcse/twenty-first-century-science-suite-combined-science-b-j260-from-2016/ (Accessed August 2021)

[2] https://www.ocr.org.uk/qualifications/as-and-a-level/chemistry-b-salters-h033-h433-from-2015/ (Accessed August 2021)

[3] Kirsti Marie Jegstad & Astrid Tonette Sinnes (2015) 'Chemistry Teaching for the Future: A model for secondary chemistry education for sustainable development', *International Journal of Science Education*, 37, (4), 655–683. DOI: 10.1080/09500693.2014.1003988 To link to this article: https://doi.org/10.1080/09500693.2014.1003988

[4] https://edu.rsc.org/eic/science (Accessed October 2021)

[5] http://science.cleapss.org.uk/Resource/Bulletin-167-Spring-2020.pdf [Membership required] (accessed October 2021)

[6] https://www.beyondbenign.org/ (Accessed October 2021)

[7] https://www.rsc.org/new-perspectives/sustainability/ (Accessed October 2021)

[8] https://www.rsc.org/new-perspectives/sustainability/sustainable-chemicals-strategy/ (Accessed October 2021)

[9] https://www.compoundchem.com/2015/09/24/green-chemistry/ (Accessed October 2021)

[10] https://www.plasticexpert.co.uk/plastic-recycling/hdpe-plastic-recycling/ (Accessed October 2021)

Postscript from Bob Worley

As stated before, we do not advocate for completely replacing the traditional practical activities that students and teachers in chemistry have carried out for many years (some for over 100 years!).

However, time marches on and we are conscious that activities in biology (DNA extraction, digestion and electrophoresis) and, to some extent, physics (use of Arduino kits and smartphones as measuring devices) have progressed to meet the modern challenges of science in the 21st century. Has this happened in chemistry?

I am so grateful to Sarah Longshaw and ASE for asking David and me to bring this book together. I am conscious of my age. I hope that David, Sarah and all the fantastic (though sometimes) under-valued technicians and teachers who enjoy teaching from observable macro events seen in Petri dishes and puddles continue to promote novel ideas in practical chemistry.

Illustrating the creativity shown by technicians in schools, I was overwhelmed by the art form of 'painting by puddles' demonstrated by Isobel Everest. So it is only fitting to show her wonderful creation of the 'Chem-eleon' made from coloured indicators in acids and alkalis.

And then, out of the blue, I received a message from the Royal Society of Chemistry. I have been awarded the 2021 Excellence in Secondary and Further Education prize. This is awarded for *'significant and sustained contributions to the development and promotion of safe practical resources for teachers worldwide, especially in the field of microscale chemistry'*. That was a real Oxide of Magnesium (OMg) moment!

So over to you. You can suggest more ideas and activities in the Google group: https://groups.google.com/g/microscalechemistry and do keep checking https://www.millgatehouse.co.uk/microscalechemistry to see if more items have been added by the authors.

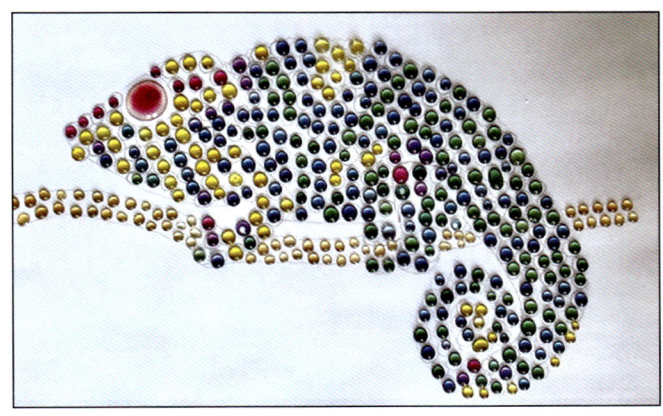

The microscale Chem-eleon by Isobel Everest